IdeaFest Journal

2025-2026

The Press at Cal Poly Humboldt

The Press at Cal Poly Humboldt
Cal Poly Humboldt Library
1 Harpst St.
Arcata, CA 95524-8299
digitalcommons.humboldt.edu

ISBN: 978-1-962081-53-5
ISSN: 2689-1891

Design: Sarah Godlin & Mia Costales
Layout: Mia Costales
Additional Editing: Mia Costales & David Fisher

The Press at Cal Poly Humboldt publishes high-quality scholarly, intellectual, and creative works by or in support of our campus community. The Press supports the Cal Poly Humboldt mission to improve the human condition and our environment by promoting understanding of social, economic, and environmental issues.

Contents

IdeaFest
Journal

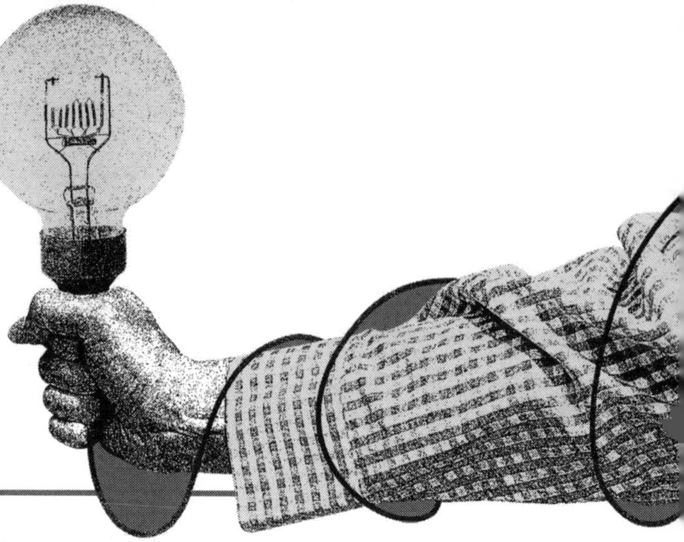

Allosteric and Uncompetitive Inhibition of Yeast Alcohol Dehydrogenase Identified by Enzyme Kinetics

Dylan J. Harrison and Jenny A. Cappuccio
Department of Chemistry & Biochemistry, Cal Poly Humboldt

Enzyme kinetics is a powerful technique used to quantify the rate at which an enzyme converts a substrate into a product and to identify potential inhibitors. In this study, kinetic analysis was employed to assess the activity of yeast alcohol dehydrogenase (YADH) in the presence and absence of sweeteners commonly found in alcoholic beverages. YADH catalyzes the oxidation of ethanol to acetaldehyde using NAD^+ as a cofactor; the reduction to NADH can be monitored spectrophotometrically at 340 nm. Based on structural similarities to the ribose moiety of NAD^+, it was hypothesized that sucrose and sucralose could directly interact with the enzyme or its cofactor, inhibiting YADH activity, resulting in a reduced rate of NADH formation. Specifically, the fructose component of sucrose was expected to interfere with the enzyme's activity. Conversely, Splenda®, being 1.1% sucralose and primarily inert maltodextrin, was predicted to have minimal impact. Michaelis-Menten and Lineweaver-Burk plots were generated using varying ethanol concentrations to determine kinetic parameters and characterize inhibition when induced. Sucrose exhibited mixed inhibition resembling noncompetitive behavior, likely binding allosterically and altering enzyme activity without interfering with substrate binding. In contrast, Splenda exhibited uncompetitive inhibition, evidenced by parallel kinetic traces and proportional decreases in both Km and Vmax. These findings demonstrate an effective approach to identifying how dietary sweeteners influence YADH activity. The results suggest that sweeteners can directly slow the enzymatic metabolism of ethanol, even in a simplified in vitro system.

Introduction

Alcohol is a central nervous system depressant that disrupts biochemical pathways by interfering with neurotransmission and metabolism. Upon ingestion, approximately 20% of ethanol is metabolized in the stomach before reaching the liver, where the majority is processed.

One effect that alcohol metabolism possesses is a reduction in the liver's ability to synthesize glucose, leading to hypoglycemia. Additionally, ethanol is an antagonist at the N-methyl-D-aspartate (NMDA) glutamate receptor, inhibiting the excitatory neurotransmitter glutamate and contributing to its depressant effects (1).

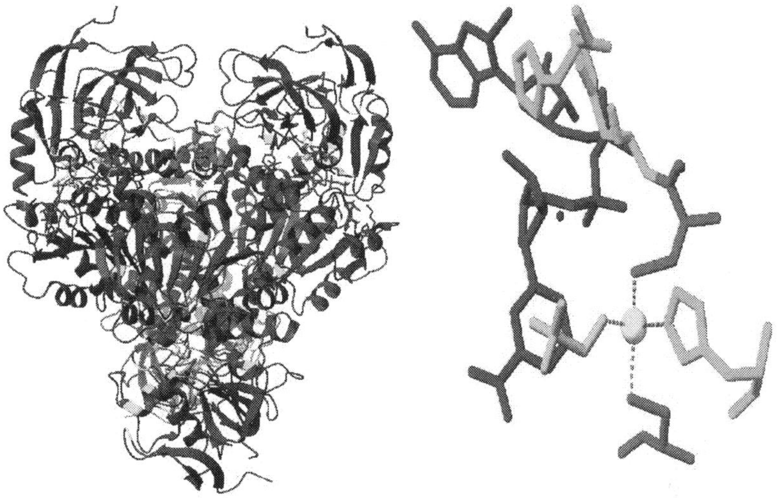

Figure 1. Crystal structure of yeast alcohol dehydrogenase (PDB: 4W6Z). The active site of YADH showing Zn^{2+} coordination. The catalytic zinc ion (yellow sphere) is tetrahedrally coordinated by two cysteines (pink), one histidine (green), and a ligand, stabilizing the transition state and orienting ethanol (cyan) for hydride transfer to NAD^+ (red). This coordination is conserved among zinc-dependent ADHs and is essential for catalysis.

The enzyme primarily responsible for ethanol metabolism is alcohol dehydrogenase (ADH), a class 1 zinc-dependent oxidoreductase. ADH catalyzes the oxidation of ethanol to acetaldehyde while reducing nicotinamide adenine dinucleotide (NAD^+) to NADH. The accumulation of NADH disrupts the NAD^+/NADH ratio, potentially inhibiting pathways such as glycolysis.

This enzyme is present across various species, including humans, fruit-eating animals, and nectar-consuming insects. Its existence suggests it is an ancient adaptation to dietary ethanol, presumed to have evolved to assist in metabolizing fermented fruits (2).

In *Saccharomyces cerevisiae*, the three zinc-containing ADH isoenzymes, ADH1, ADH2, and ADH3, serve different physiological roles. ADH1 is the primary enzyme in ethanol fermentation, ADH2 is glucose-repressed and functions in ethanol oxidation, and ADH3 localizes to mitochondria for redox balance during respiration (3). Each subunit contains two zinc ions: a structural zinc for protein stability and a catalytic zinc that facilitates ethanol binding along with hydride transfer to NAD^+ (4). This reaction produces NADH, which can be monitored spectrophotometrically at 340 nm due to its characteristic absorbance.

Acetaldehyde is subsequently oxidized by aldehyde dehydrogenase (ALDH), producing more NADH and acetate. Acetate is then converted into acetyl-CoA in an ATP-dependent process. Because these reactions generate NADH, tracking absorbance at 340 nm provides a direct method for measuring ADH activity.

This experiment was inspired by a study conducted at Northern Kentucky University, which found that individuals who consumed alcohol with diet soda had significantly higher peak blood alcohol concentrations (BACs) compared to those who drank alcohol with regular soda (1). The researchers proposed that non-nutritive sweeteners like sucralose accelerate gastric emptying, leading to faster ethanol absorption. However, this observation raises a biochemical question: could sweeteners influence alcohol metabolism by interacting directly with ADH?

To explore this, two sweeteners were selected for an in vitro investigation: glucose, a natural monosaccharide metabolized through glycolysis, and sucralose, a non-caloric artificial sweetener. Sucralose is a chlorinated analog of sucrose and is the active ingredient in Splenda® (1.1% sucralose with maltodextrin, a filler). Unlike glucose, sucralose is not metabolized in humans and does not contribute to NADH production. However, due to its structural similarity to sucrose, it may still bind to or affect ADH through non-classical inhibition mechanisms.

Glucose, a primary energy source, is metabolized via glycolysis, generating NADH and pyruvate. At high concentrations, glucose may shift the NAD^+/NADH balance, thereby reducing the availability of NAD^+ required for ADH activity and possibly slowing ethanol metabolism. On the other hand, sucralose is an artificial sweetener, the primary component of Splenda, and is not metabolized by human cells. Structurally, sucralose resembles sucrose but features chlorine atoms in place of hydroxyl groups, rendering it non-reactive in glycolytic or oxidative pathways. Its interaction with ADH, if any, would likely occur via direct binding or inhibition rather than through metabolic effects (4).

This project investigates whether sweeteners inhibit yeast ADH activity and characterize the nature of that inhibition. The experiment is conducted in vitro under cell-free conditions using NADH production to determine the rate of enzyme activity. Systemic physiological effects are excluded in this system, but direct interactions between sweeteners and ADH may still yield measurable changes in enzyme kinetics. Observed inhibition could be beneficial to nutritionist.

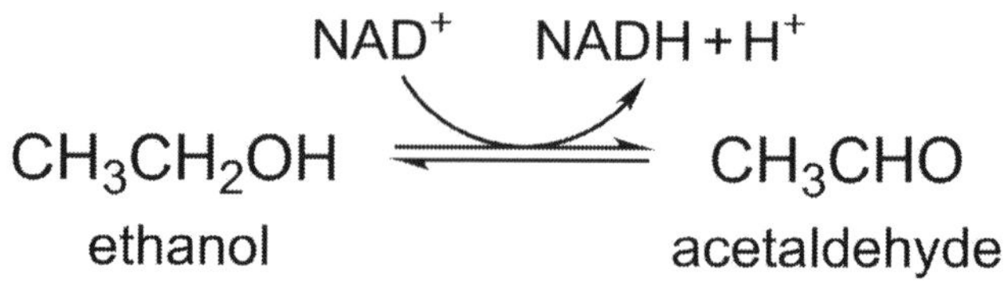

Figure 2. YADH catalyzes the oxidation of ethanol to acetaldehyde, using NAD^+ as a cofactor and producing NADH and H^+ as byproducts.

Methods

The ethanol used (CAS # 64-17-5), DSP-MD. 43, was anhydrous, meeting USP Specs and acquired from Decon Laboratories, Inc. The alcohol dehydrogenase was from yeast and purchased through Tokyo Chemical Industry Co. (TCI) product number A0200. The vial of β-Nicotinamide adenine dinucleotide (NAD+) used was from Sigma-Aldrich. The biochemistry stockroom at Cal Poly Humboldt provided monobasic potassium phosphate and sucrose.

Reagent Preparation

- 50 mM Phosphate Buffer (pH 7.4). Using an Erlenmeyer flask, 1.70 grams of monobasic potassium phosphate (KH2PO4) was weighed and dissolved in about 200 mL of Milli-Q. Using a stir plate and a calibrated pH probe, the solution was titrated to a pH of 7.4 with 1 M NaCl, with caution near pH 7, as small additions make big changes. Once the desired pH was reached, the final volume was brought up to 250 mL using Milli-Q water. All further reagents were prepared using this buffer to maintain experimental consistency.
- NAD+ 10 mM. In a Falcon tube, 0.0332 grams of NAD+ was weighed and dissolved with 5 mL of the 50 mM phosphate buffer, pH 7.4. This solution was prepared when needed.

- Yeast Alcohol Dehydrogenase 1 mg/ 1mL.
- In a Falcon tube, 5 mg was weighed and 5 mL of 50 mM phosphate buffer added to dissolve the enzyme.
- 80 proof, 40% Ethanol. Diluted 200 proof ethanol to 80 proof using the 50 mM phosphate buffer.
- 100 mM Sucrose. In a Falcon tube, 0.3428 grams of sucrose was weighed out and dissolved with 10 mL of 50 mM phosphate buffer.
- 5.52 mM Sucralose. Splenda is a mixture consisting of 1.1% sucralose and the rest being a filler, maltodextrin. Calculations for solving the molarity of sucralose are shown in the Appendix. In a Falcon tube, 1.9006 grams of Splenda was weighed and dissolved with 10 mL of 50 mM phosphate buffer. A volume of 5 mL was initially used to make 10 mM, but it did not dissolve, requiring an additional 5 mL.

Procedure

All assays were performed using quartz cuvettes and an Agilent HP8453 Diode Array Spectrophotometer with UV-visible software. Initial rates were determined from data collected between 5-60 seconds.

Part A. Measurement of YADH concentration

To determine the specific activity of the enzyme, the exact concentration of the enzyme must be known. The wavelength of the spectrophotometer was set to 280 nm, buffer was pipetted into a quartz cuvette to blank the instrument. Fill a quartz cuvette to the optical window with the YADH and read the absorbance at 280 nm. The concentration of the enzyme may be determined by using Beer's Law, $A = \epsilon cl$, the extinction coefficient at 280 nm is 14.6. Concentration was converted to mg of YADH/ mL.

Kinetic Assays of Ethanol

Solutions of NAD+ and YADH were stored on ice, while the ethanol and phosphate were kept at room temperature. All reagents were placed in the fridge at the end of each day. The spectrophotometer was turned on and warmed up for 15 minutes before measuring the absorbance. Table 1 shows the volume ratios put into cuvettes for these runs. A heated cell holder was utilized to incubate the cuvette at a temperature of 37 °C before initiating the reaction with the enzyme. On the Agilent, the kinetics mode was used to determine the initial rate of reaction. The program utilized had a wavelength of 340 nm with a runtime of 180 seconds and a cycle time of 5 seconds. This program carried out a kinetics run where a spectrum at A340 was measured every 5 seconds for 3 minutes. The instrument was blanked with 50 mM 3 phosphate buffer. Pipetted buffer, NAD+, and ethanol into the quartz cuvette, placed parafilm over, and inverted to mix, then incubated at 37 °C for 5 minutes. After incubation, initiation of the reaction was done by pipetting 50 µL of ADH, mixing by covering with parafilm and inverting about three times, then immediately placing the cuvette in the spectrophotometer and clicking sample to begin collecting rate data. This method was repeated for assays 1-6 of Table 1.

Parameters Determination

Table 3 in the results was created, which displays the ΔA/min (rate vs enzyme concentration, mg/assay). The units were converted to a more useful form, µmole/min, the appendix shows this math. Using the table created, a plot of rate vs substrate concentration was made to generate a Michaelis-Menten plot (initial velocity versus substrate concentration). Inverses of the rate and concentration were plotted to create a Lineweaver-Burk graph to determine the Km and Vmax of the substrate, ethanol.

Inhibition of Substrate

To determine if sweeteners inhibit the activity of enzyme turnover, the amount of inhibitor induced was kept constant while the substrate concentration varied. Table 2 shows the ratios of reagents pipetted into the cuvette for rate determination.

Table 1. Rations of varying amounts of substrate to determine the Km and Vmax measured at 340 nm.

Assay:	1	2	3	4	5	6
Phosphate Buffer, pH 7.4	749 µL	747 µL	744 µL	741 µL	738 µL	735 µL
Ethanol (6.84 M) 40%, 80 Proof	1 µL	3 µL	6 µL	9 µL	12 µL	15 µL
Nicotinamide adenine dinucleotide (NAD+)	200 µL	200 µL	200 µL	200 µL	200 µL	200 µL
Yeast Alcohol Dehydrogenase (YADH)	50 µL	50 µL	50 µL	50 µL	50 µL	50 µL

Table 2. Ratios used for the substrate inhibition runs consisted of a constant volume of inhibitor with varying substrate, measured at 340nm.

Assay:	1	2	3	4	5	6
Phosphate Buffer, pH 7.4	739 µL	737 µL	734 µL	731 µL	728 µL	725 µL
Ethanol (6.84 M) 40%, 80 Proof	1 µL	3 µL	6 µL	9 µL	12 µL	
Nicotinamide adenine dinucleotide (NAD+)	200 µL	200 µL	200 µL	200 µL	200 µL	200 µL
Yeast Alcohol Dehydrogenase (YADH)	50 µL	50 µL	50 µL	50 µL	50 µL	50 µL
Inhibitor (Sucrose/ Splenda)	10 µL	10 µL	10 µL	10 µL	10 µL	10 µL

Solutions were prepared in the same manner as those in part B. The instrument was blanked with 50 mM phosphate buffer. Each assay had all of the reagents except the enzyme pipetted, then incubated for 5 mins at 37 °C using the thermostatted cell holder attached to the spectrophotometer. After incubation, 50 μL of YADH was added to the cuvette, parafilmed, and inverted to mix. Once the enzyme was added, the cuvette was placed into the cell to begin measurement of NADH formation. Both inhibitors, sucrose and Splenda, had initial rates measured in the same way.

Measured *ΔA*/min were converted to μmole/min and then plotted against substrate concentration. The plotted data produced a hyperbolic line (Michaelis-Menten plot). The inverses of rate (μmole/min) vs [S] were plotted to linearize the data, making a Lineweaver-Burk plot.

Results

Michaelis-Menten plots were generated to illustrate the relationship between varying concentrations of ethanol and the initial rate of NADH formation, catalyzed by yeast alcohol dehydrogenase (YADH). Figure 3 shows the plots of the uninhibited and inhibited assays overlaid. A typical hyperbolic curve for the control reaction exhibited normal enzyme saturation behavior. Inhibition with sweeteners was performed with sucrose and Splenda, which assessed how these additives influence activity. While Michaelis-Menten curves indicated a decrease in reaction rate in the presence of both inhibitors, the exact mode of inhibition could not be conclusively determined from these plots alone. To characterize inhibition type, Lineweaver-Burk plots were constructed to linearize the kinetic data and enable quantitative analysis (Figure 4). This allowed for comparison of changes in substrate affinity and maximum reaction rate, key components to identifying the type of inhibition.

In the presence of sucrose, the Lineweaver-Burk plot (Figure 5) showed intersecting lines between the control and inhibited reactions. This was accompanied by a decrease in Vmax (from 0.0112 μmol/min to 0.00866 μmol/min) with a relatively small increase in Km (from 64.84 mM to 80.22 mM). This pattern is characteristic of mixed inhibition, closely resembling noncompetitive behavior. It suggests that sucrose binds to an allosteric site on YADH, altering enzyme conformation and reducing catalytic efficiency without strongly interfering with ethanol binding. The structural similarity between sucrose and the ribose component of NAD^+ may contribute to this interaction, although sucrose lacks the redox-ac-

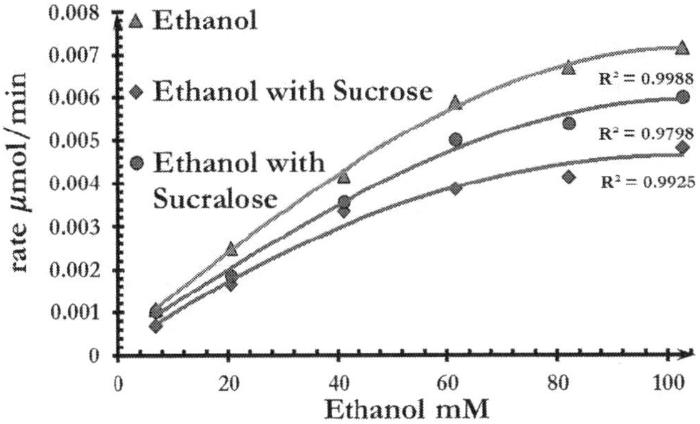

Figure 3. Michaelis-Menten plot showing the relationship between ethanol concentration and the initial rate of NADH formation catalyzed by yeast alcohol dehydrogenase. The green curve represents the uninhibited control, while the red and purple curves show the effects of sucrose and Splenda, respectively. Both sweeteners caused a decrease in the reaction rate. Data points were fit to a hyperbolic curve, and R^2 values indicate strong correlation with the Michaelis-Menten model.

tive phosphate and nitrogenous groups needed to participate directly in catalysis.

In contrast, when Splenda was added as the inhibitor, the Lineweaver-Burk plot (Figure 6) revealed parallel lines relative to the control, indicating uncompetitive inhibition. Both Km and Vmax decreased proportionally (Km = 44.19 mM, Vmax = 0.0074 µmol/min), an indication that sucralose binds exclusively to the enzyme-substrate complex. This form of inhibition locks the enzyme in an inactive state after ethanol is bound, preventing product formation. The chlorinated structure of sucralose, a modified derivative of sucrose, likely enhances its chemical stability and allows it to mimic aspects of the bound substrate without participating in catalysis.

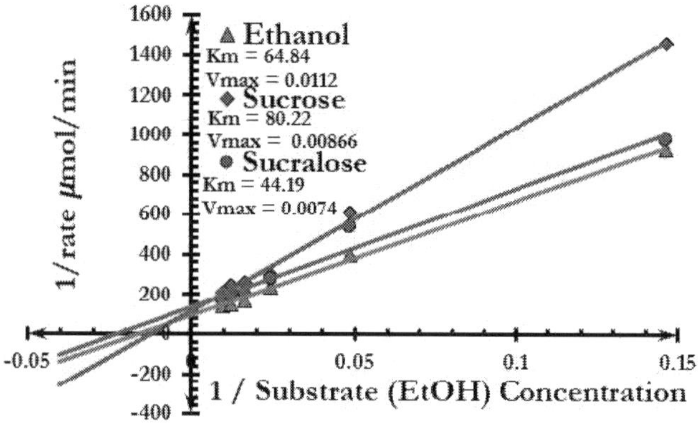

Figure 4. Lineweaver-Burk plots displaying the kinetic effect on YADH activity. The control exhibited a Km of 64.84 mM and a Vmax of 0.0112 µmol/min. Sucrose induction produced a steeper slope and higher y-intercept, with a Km of 80.22 mM and Vmax of 0.00866 µmol/min, consistent with mixed inhibition resembling noncompetitive behavior. When induced with Splenda, a parallel kinetic trace was generated, with Km and Vmax reduced to 44.19 mM and 0.0074 µmol/min, respectively. This trend confi rms uncompetitive inhibition, where the inhibitor binds only to the enzyme-substrate complex, reducing catalytic efficiency without affecting substrate binding affi nity.

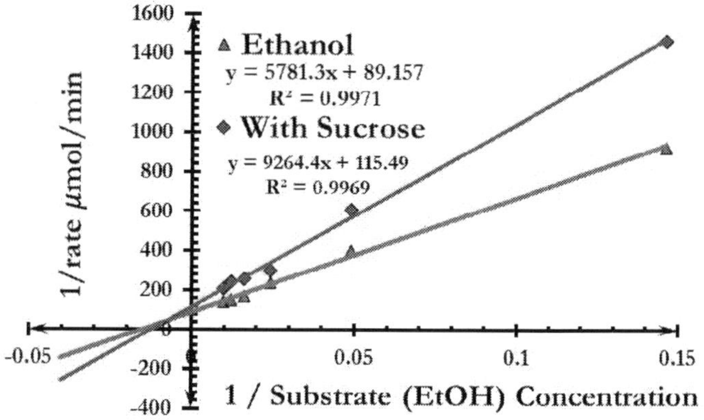

Figure 5. This plot displays the double-reciprocal relationship between ethanol concentration and the initial rate of NADH formation, comparing uninhibited enzyme activity to the reaction in the presence of 1 mM sucrose. The presence of sucrose results in a steeper slope and an increased y-intercept, indicating a reduction in Vmax. A slight shift in the x-intercept suggests a small increase in Km. The observed kinetic changes are consistent with mixed inhibition, where sucrose interacts with both the free enzyme and the enzyme-substrate complex, likely by binding to an allosteric site and altering catalytic efficiency without fully preventing substrate binding.

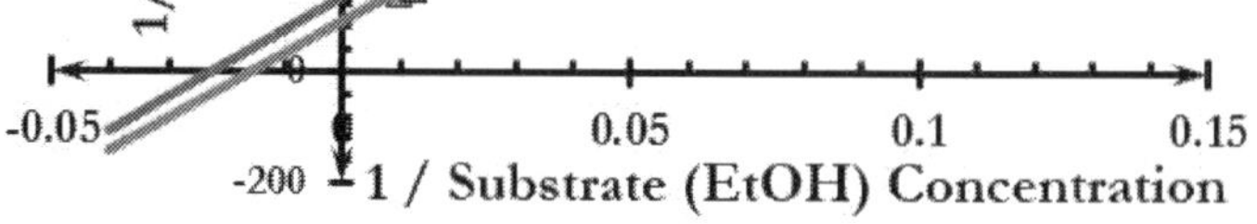

Figure 6. This overlaid Lineweaver-Burk plot highlights the eff ect of Splenda on the enzymatic oxidation of ethanol by YADH. In the presence of Splenda, the data produced a parallel line relative to the uninhibited reaction, with a proportional decrease in both Km and Vmax. This pattern is characteristic of uncompetitive inhibition, in which the inhibitor binds exclusively to the enzyme-substrate complex. The sucralose component of Splenda likely stabilizes this complex in an inactive state, reducing overall catalytic

These findings are summarized in Figure 4, demonstrating the inhibition patterns that sucrose and sucralose induce on YADH activity. The results discovered that sucrose acts as a noncompetitive inhibitor, likely binding to an allosteric site, while Splenda exhibits uncompetitive inhibition, interfering only after substrate binding. Exploring how natural and artificial sweeteners can modulate enzyme activity in distinct ways depending on their structure and binding behavior.

To quantitatively assess the efficiency of YADH, the enzyme's turnover number (kcat) and catalytic efficiency (kcat/Km) were calculated from the uninhibited ethanol assay. With a measured Vmax of 0.0112 μmol/min (1.87×10^{-10} mol/s) and an enzyme concentration of 3.40×10^{-8} mol/L, the resulting kcat was 5.50 s^{-1}, indicating that each enzyme molecule catalyzes approximately 5.5 ethanol-to-acetaldehyde conversions per second under saturated conditions. Using the determined Km of 64.84 mM, the catalytic efficiency was calculated as 84.8 $M^{-1}s^{-1}$. These values establish a functional baseline for YADH activity and provide essential context for evaluating the extent of inhibition caused by sucrose and Splenda. The relatively modest catalytic efficiency suggests that YADH is not diffusion-limited, and its activity can be meaningfully altered by structural inhibitors, as demonstrated in subsequent assays.

Table 3. Rate of NADH Formation Uninhibited (Ethanol only)

Assay	[S] mM	ΔA/min	μmol/min
1	6.84	0.0067	1.08×10^{-3}
2	20.52	0.0155	2.50×10^{-3}
3	41.04	0.0259	4.17×10^{-3}
4	61.56	0.0366	5.88×10^{-3}
5	82.08	0.4152	6.67×10^{-3}
6	102.6	0.4439	7.14×10^{-3}

Table 4. Rate of NADH Formation Inhibited with Sucrose, 10 μL Induction

Assay	[S] mM	ΔA/min	μmol/min
1	6.84	0.0042	6.81×10^{-4}
2	20.52	0.0102	1.64×10^{-3}
3	41.04	0.0207	3.33×10^{-3}
4	61.56	0.0239	3.85×10^{-3}
5	82.08	0.0256	4.12×10^{-3}
6	102.6	0.0297	4.79×10^{-3}

Table 5. Rate of NADH Formation Inhibited with Splenda, 10 μL Induction

Assay	[S] mM	ΔA/min	μmol/min
1	6.84	0.0063	1.02×10^{-4}
2	20.52	0.0115	1.85×10^{-3}
3	41.04	0.0220	3.56×10^{-3}
4	61.56	0.0309	4.99×10^{-3}
5	82.08	0.0332	5.36×10^{-3}
6	102.6	0.0371	5.98×10^{-3}

Discussion

The goal of this experiment was to investigate how different sweeteners influence the enzymatic activity of yeast alcohol dehydrogenase, an enzyme that is responsible for oxidizing ethanol to acetaldehyde using NAD^+ as a cofactor. Through kinetic assays and analysis using Michaelis-Menten and Lineweaver-Burk plots, the effects of sucrose and Splenda (sucralose) on ethanol metabolism were evaluated and inhibition patterns were characterized. The results revealed that sucrose functions as a mixed inhibitor, closely resembling noncompetitive inhibition. The presence of sucrose led to a significant decrease in Vmax 7 and a slight increase in Km, suggesting that it binds to a site other than the active site, interfering with enzyme function but still allowing ethanol to bind. This behavior aligns with sucrose's molecular structure, despite not being redox-active, and also shares similarities with the ribose component of NAD^+. These similarities may allow sucrose to bind to regulatory or allosteric sites on YADH, causing conformational changes that reduce catalytic efficiency without fully blocking substrate access.

In contrast, Splenda, a commercial artificial sweetener composed primarily of maltodextrin and 1.1% sucralose, exhibited uncompetitive inhibition. In this case, both Km and Vmax decreased proportionally, consistent with an inhibitor that binds only after the substrate is already attached to the enzyme. This suggests that sucralose stabilizes the enzyme-substrate complex in an inactive state, preventing turnover. Structurally, sucralose is a chlorinated analog of sucrose, rendering it chemically inert and resistant to enzymatic breakdown. Its resemblance to carbohydrate structures likely enables it to interact with the enzyme once the substrate is bound, mimicking aspects of the transition state or product complex.

These findings are particularly relevant in the context of alcohol consumption with mixers. Both natural and artificial sweeteners were found to slow the rate of ethanol metabolism by inhibiting YADH, but through different mechanisms. Notably, uncompetitive inhibitors like sucralose may be more effective at reducing ethanol breakdown when alcohol is already present, potentially prolonging intoxication or affecting blood alcohol concentrations. This raises interesting questions about how beverage formulations might influence metabolic processing and individual responses to alcohol.

Overall, the experiment demonstrated the utility of enzyme kinetics in identifying and characterizing different modes of inhibition. Lineweaver-Burk analysis allowed for clear differentiation between inhibition types and provided insight into how molecular structure governs enzyme-inhibitor interactions. Future experiments could further explore dose-dependent inhibition curves and test pure sucralose independent of maltodextrin or pure glucose and fructose rather than sucrose. These findings enhance our understanding of YADH regulation and highlight how common dietary additives can modulate enzyme behavior in ways that may affect human metabolism.

Conclusion

This experiment demonstrated that sweeteners inhibit yeast alcohol dehydrogenase activity through distinct mechanisms, reflecting their structural and functional differences. Sucrose acted as a mixed inhibitor, primarily noncompetitive, reducing catalytic efficiency by binding allosterically and altering enzyme conformation without strongly affecting substrate binding. In contrast, Splenda, composed mainly of maltodextrin and sucralose, exhibited uncompetitive inhibition, binding only after the substrate was engaged and effectively trapping the enzyme-substrate complex in an inactive state.

These mechanistic differences had clear effects on kinetic parameters: sucrose caused a decrease in Vmax with a slight increase in Km, while Splenda led to a proportional decrease in both, consistent with classic uncompetitive inhibition. Beyond the lab bench, these results carry real-world implications. They suggest that consuming alcohol with diet mixers containing sucralose could suppress ethanol metabolism more significantly than sugar-based mixers, potentially leading to elevated and prolonged blood alcohol concentrations. By combining kinetic modeling with molecular insight, this study highlights how common dietary additives can influence the enzymatic processing of alcohol. It underscores the broader impact of sweetener choice on alcohol metabolism, an intersection of nutritional chemistry and public health that warrants further attention.

References

1. Wu, C. D.; Brick, L. A.; Carr, R. D. Effects of Regular and Diet Soda on Alcohol Absorption: A Laboratory Study. Drug Alcohol Depend. 2016, 160, 164–167. https://doi.org/10.1016/j.drugalcdep.2015.10.015
2. Dudley R. Fermenting fruit and the historical ecology of ethanol ingestion: Is alcoholism in modern humans an evolutionary hangover? Addiction. 2002 Apr;97(4):381-8. doi: 10.1046/j.1360-0443.2002.00002.x. PMID: 11964055.
3. Leskovac, V.; Trivić, S.; Peričin, D. The Three Zinc-Containing Alcohol Dehydrogenases from Baker's Yeast, Saccharomyces cerevisiae. FEMS Yeast Res. 2002, 2 (4), 481–494. https://doi.org/10.1111/j.1567-1364.2002.tb00116.x
4. Raj, S. B.; Ramaswamy, S.; Plapp, B. V. Yeast Alcohol Dehydrogenase Structure and Catalysis. ACS Biochemistry 2014, 53, 36, 5791–5803 https://doi.org/10.1021/bi5006442
5. Dudley, R.; Maro, A.; Human Evolution and Dietary Ethanol. Nutrients 2021, 13, 2419. https://doi.org/10.3390/nu13072419
6. Nelson, David L., et al. Lehninger Principles of Biochemistry. Macmillan Learning, 2021.

Appendix

Rate (ΔA/min) vs enzyme concentration (mg per assay) conversion to μmole/min, a more useful unit. Units are converted to determine the amount of product formed per unit of time. Beer's law: $c = A/\epsilon l$, where A is the absorbance change that occurs per minute, l is the light path length through the cuvette (1 cm), ϵ is the molar absorption coefficient for the product, (at 340 nm NAD+ molar absorptivy is 6220 M-1cm-1) and c is the concentration of the product. The absorbance change during 1 minute for the production of 1 molar NAD+ is 6220:

1 molar = A / (6220 $M^{-1}cm^{-1}$ x 1 cm)   A = 6220

To convert the raw data to moles of product formed per minute per liter, each ΔA/min was divided by 6220.

ex. Uninhibited assay 3

$\Delta c = \Delta A/\ min\ (\epsilon l) = 0.0259/min \div (6220\ M^{-1}cm^{-1}\ x\ 1\ cm) = 4.17\ x10^{-6}\ M/min$

Now to moles/min:

$4.17\ x10^{-6}\ moles/liter\ min\ x\ .001\ liters = 4.17\ x10^{-9}\ M/min$

Now convert to μmoles/min:

$\Delta c = 4.17\ x10^{-2}\ \mu moles/min$

Splenda is 1.1% sucralose with the remainder being a filler, maltodextrin. To account for this, we must follow these steps. Sucralose has a molecular weight of 397.64 g/mol.

$1.9006\ g\ Splenda\ x\ .011 = 0.0209066\ g\ sucralose$

$0.0209066\ g\ sucralose\ x\ (1\ mol/397.64\ g\ sucralose) = 5.2577\ x10^{-5\,M}$

$5.2577\ x10^{-5}\ mol/L\ x\ (1\ L/.01\ L) = 5.2577\ x10^{-3M}$

$5.2577\ x10^{-3}\ M\ x\ (1000mM/1\ M) = 5.26\ mM$

IdeaFest
Journal

Exploring the Relationship Between Tree DBH and Lichen Cover on Coast Redwood

Grace Personeus, Sierra Gittleman and Alisa Zarattini
Department of Environmental Science & Management, Cal Poly Humboldt

We investigated the relationship between Diameter at Breast Height (DBH) and lichen cover on coast redwood (*Sequoia sempervirens*) to see if larger trees had higher percentages of cover. We measured cover using 10 cover classes on the southeastern and southwestern sides of trees in a secondary growth forest. We found a significant relationship between DBH and cover, yet no significant difference between the two sides measured. We suggest that further research with a larger sample size and more varied age classes will provide further insight into the relationship.

Introduction

Lichens are unique organisms. They are capable of anhydrobiosis, which allows them to survive while losing more than 99% of their water content (Armaleo & Chiou, 2021). Due to this ability, many lichens can survive even after desiccation for months (Kranner et. al 2008). Old-growth coast redwoods (*Sequoia sempervirens*) are amongst the oldest organisms on the planet and are rich in lichen diversity (O'Hara et al., 2017). In an epiphyte survey performed on nine large coast redwoods, 137 different species of lichens were found (Bendiksby et al., 2018). We decided to investigate if there is a relationship between lichen cover and tree DBH, since size is often a proxy for age, and both organisms are long-living. We believe there is a significant relationship between lichen cover and DBH because larger trees are older and allow more time for lichen colonization to occur.

Methods

We used a random sampling method to sample every other tree along Jolly Giant Creek in the Arcata Community Forest until we reached a sample size of 15. The sampling method and sample size was decided based on the time allocated to complete our study. We measured DBH and lichen cover class of each sampled tree. We chose this location because it is a secondary growth forest mainly composed of coast redwood of varying DBH.

Due to the presence of indistinguishable microlichen, we decided to measure the percent cover of all lichen species present in our quadrats. We placed 10x10-inch quadrats under the two-foot mark on the southeastern and southwestern sides of trees to determine cover (Figure 1). We then used 10 classes to determine cover: 1 - 0-9%, 2 - 10-19%, 3 - 20-29%, 4 - 30-39%, 5 - 40-49%, 6 - 50-59%, 7 - 60-69%, 8 - 70-79%, 9 - 80-89%, 10 - 90-100%. We measured at the two-foot mark because we found it was relatively undisturbed from human activity that strips bark, like slacklining and hammocking. We then conducted two statistical tests to analyze our data. First, we ran a regression analysis to assess the relationship between lichen cover and DBH on both sides of the trees measured. We also ran a paired t-test to see if there was a difference in cover between the southeastern and southwestern sides of the same trees.

Figure 1: We used a 10x10-inch quadrat placed under the two-foot mark on both sides of the coast redwood trees (Sequoia sempervirens) to aid in determining the cover class of all lichen species present.

Results

Of the 15 coast redwoods measured, the dominant lichen types we saw were crustose and squamulose. The regression analysis we ran indicates a significant relationship between DBH and cover ($R2 = 0.14$, $P = 0.04$; Figure 2). The paired t-test we ran shows no significant difference in cover between the two sides measured ($P = 0.63$; Figure 3).

Discussion

We found a significant relationship between DBH and lichen cover on coast redwood ($P = 0.04$; Figure 2). However, we found no significant difference between the two sides measured ($P = 0.63$; Figure 3). Previous research shows that species like Douglas fir (*Pseudotsuga menziesii*) and African locust bean (*Parkia biglobosa*) have demonstrated increased lichen cover on larger trees

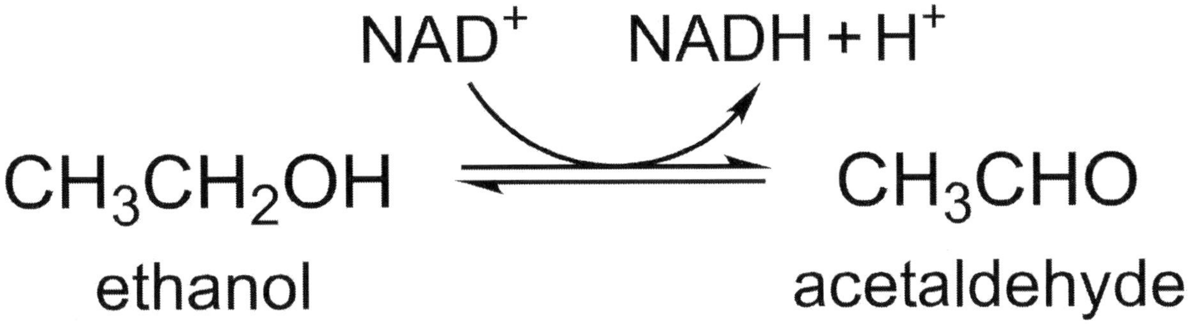

Figure 2: Lichen cover determined by 10 cover classes (0-100%) on two sides of coast redwood (Sequoia sempervirens) trees along Jolly Giant Creek in the Arcata Community Forest, Humboldt County, CA (P = 0.04).

(McCune & Antos, 1982; Buba & Danmallam, 2019). Other research shows that species like grand fir (Abies grandis) have decreased cover with increased size (McCune & Antos, 1982). Suggested reasons for varying cover with size are attributed to factors influencing growth, such as canopy structure and bark texture, due to their impact on light and moisture levels (Rosentreter, 1990). In our sample, there is a high level of variation in cover on both sides, which is likely due to varying levels of light, moisture, and disturbance present in the ecosystem (Figure 2). Although we found a statistically significant relationship between DBH and lichen cover, the low R2 value of 0.14 indicates a weak relationship between DBH and lichen cover. Lichen cover is likely more affected by canopy structure and bark texture than DBH alone (Rosentreter, 1990).

One consideration for possible error in our study is incorrect lichen cover estimation. Most of the lichen cover consisted of unevenly dispersed microlichen, making it difficult to assess the percent cover. A limitation could be the quadrat location since some individuals exhibited a high cover class on other parts of the tree but not where we decided to measure. The location of our study

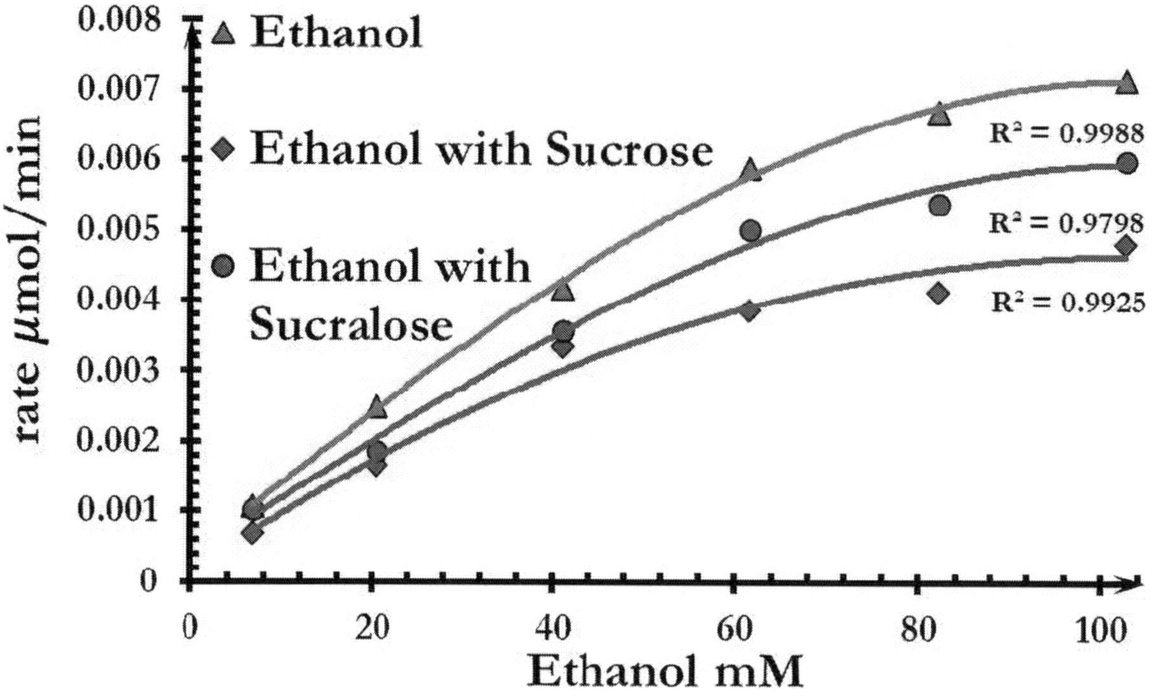

Figure 3: Difference in lichen cover based on side of tree in coast redwoods (Sequoia sempervirens) along Jolly Giant Creek in the Arcata Community Forest, Humboldt County, CA (P = 0.63).

could be another limitation. No old-growth was present causing us to only measure relatively young trees present in a secondary forest instead of older trees. This could be largely influential as age may play a role in lichen colonization (Uliczka & Angelstam, 1999).

Though we found a significant relationship between DBH and lichen cover, a larger sample size with more varied age classes would provide further insight into the relationship present.

Additionally, our goal with sampling every other tree along Jolly Giant Creek was to use a random sampling method, but this may have impacted our results. Sampling trees directly off the trail due to convenience and time constraints likely resulted in edge effects as sun exposure affects lichen cover. Measuring a variety of trees at different positions within a stand could have better accounted for other factors that affect lichen cover. Ultimately, another study is needed to explore the relationship between redwood tree DBH and lichen cover. We hope our study can provide insightful consideration to further research.

References

Armaleo, D., & Chiou, L. (2021). Modeling in yeast how rdna introns slow growth and increase desiccation tolerance in lichens. G3 Genes|Genomes|Genetics, 11(11). https://doi.org/10.1093/g3journal/jkab279

Bendiksby, M., Reese Næsborg, R., & Timdal, E. (2018). Xylopsora canopeorum (Umbilicariaceae), a new lichen species from the canopy of Sequoia sempervirens. MycoKeys, 30, 1–15. https://doi.org/10.3897/mycokeys.30.22271

Buba, T., & Danmallam, B. (2019). Effects of Tree Size and Bark Roughness of Parkia Biglobosa on Lichen Colonization in Amurum Forest Reserve: Implication for Conservation. Science Forum (Journal of Pure and Applied Sciences, 17(1), 73-83. http://dx.doi.org/10.5455/sf.32286

Kranner, I., Beckett, R., Hochman, A., & Nash, T. H. (2008). Desiccation-Tolerance in Lichens: A Review. The Bryologist, 111(4), 576–593. https://doi.org/10.1639/0007-2745-111.4.576

McCune, B., & Antos, J. A. (1982). Epiphyte Communities of the Swan Valley, Montana. The Bryologist, 85(1), 1–12. https://doi.org/10.2307/3243136

O'Hara, K. L., Cox, L. E., Nikolaeva, S., Bauer, J. J., & Hedges, R. (2017). Regeneration Dynamics of Coast Redwood, a Sprouting Conifer Species: A Review with Implications for Management and Restoration. Forests, 8(5), 144. https://doi.org/10.3390/f8050144

Rosentreter, R. (1990). Indicator value of lichen cover on desert shrubs. Biological Conservation, 59(2), 273-273. https://www.fs.usda.gov/rm/pubs_int/int_gtr276/int_gtr276_282_289.pdf

Uliczka, H., & Angelstam, P. (1999). Occurrence of Epiphytic Macrolichens in Relation to Tree Species and Age in Managed Boreal Forest. Ecography, 22(4), 396–405. https://doi.org/10.1111/j.1600-0587.1999.tb00576.x

IdeaFest
Journal

Interrogating Settler Narratives of Erasure in the Schatz Tree Farm Archives

Sangeeta Sarkar
Department of Native American Studies, Cal Poly Humboldt

In 1987, L. W. Schatz donated a 385-acre parcel of land to Cal Poly Humboldt to be used as an experimental station, outdoor classroom, and demonstration area for forestry educators and students. Previous student research and other institutional narratives have extensively documented the history of this land from a settler perspective. However, this documentation contains few mentions of the Indigenous history of the landscape. Through literature review and archival research, this paper critically analyzes settler colonial narratives around the Schatz Tree Farm and the ways in which they contribute to Indigenous erasure. The predominant forms of narrative erasure in settler colonial histories of the Tree Farm include Patrick Wolfe's logic of elimination, historically locating Indigenous peoples in the past tense, and the minimization of Traditional Ecological Knowledge (TEK). This paper argues that in order for Cal Poly Humboldt to actualize its stated commitments to decolonization, the university must return the Schatz Tree Farm to Indigenous stewards.

Introduction

We acknowledge that the L.W. Schatz Demonstration Tree Farm is located on the unceded lands of the Whilkut people, where they have resided from time immemorial. We encourage all to gain a deeper understanding of their history and culture. As an expression of our gratitude we are genuinely committed to developing trusting, reciprocal, and long lasting partnerships with the Whilkut people as well as all of our neighboring tribes. (*Whilkut Tribe handout*, 2023, para. 1)

My friend Jade doesn't believe in bank accounts, so they keep their entire savings in a wallet. It's not just any wallet, either; it's an heirloom that's been passed down through generations of their family, carefully making its way from the care of one ancestor to the next. Imagine that one day, I decided to steal the wallet from Jade. Through a combination of trickery and violence, I wrested the wallet out of their hands and slipped it into my pocket. I decided to keep the wallet and claim it as my own, along with all the cash inside of it. Over the years, I invested Jade's life savings and collected dividends, getting richer and richer while Jade was left to rebuild their life from scratch. Many years laters, I started feeling a little guilty about the theft and decided I should let everyone know the wallet was actually stolen. I sat down to write a wallet acknowledgement, which goes something like this:

> *"I acknowledge that Sangeeta's Wallet is the unceded wallet of Jade. The wallet is an heirloom, passed down in Jade's family for generations. I encourage everyone to gain a deeper understanding of Jade's history and culture. As an expression of my gratitude, I am genuinely committed to developing a trusting, reciprocal, and long lasting partnership with Jade."*

Notably, this wallet acknowledgement says nothing about me returning the wallet or attempting to repair the harm its theft had caused Jade. In all honesty, I plan to keep the wallet and balk at the thought of returning it; after all, I'm a far better caretaker of this wallet than Jade ever was. What were they even doing with all that money anyway? No, I should hold on to it for safekeeping. Given this scenario, you may ask, how could Jade even begin to trust me when Irefuse to return what has been stolen? It logically follows that the first step to developing this "trusting, reciprocal, and long lasting partnership" I committed to in the wallet acknowledgement has to be to return the wallet and cash to Jade.

Now imagine it wasn't a wallet that was stolen, but homelands; in this case, the Whilkut land that would come to be known as the L. W. Schatz Demonstration Tree Farm[1] The wallet acknowledgement presented earlier is a variation on the Schatz Tree Farm's land acknowledgement. In this paper, I interrogate the ways in which settler colonialism shows up in Cal Poly Humboldt's documented history of the Schatz Tree Farm, including the land acknowledgement and other university narratives. I end this chapter with an argument for Cal Poly Humboldt to return the Schatz Tree Farm to the Hoopa Valley Tribe and/or the Blue Lake Rancheria, as it is located within their ancestral territories. Giving this #LandBack[2] would demonstrate the genuine commitment "to developing trusting, reciprocal, and long lasting partnerships" that is stated in the land acknowledgement (*Whilkut Tribe handout*, 2023, para. 1).

Background

Settler Colonial History of U.S. Universities

Cal Poly Humboldt's land ownership is situated within U.S. universities' long history of settler colonial violence against Indigenous peoples, including dispossession of Indigenous lands and knowledges. Many settler universities are physically located on unceded Indigenous lands,

1 This oversimplification is made for illustrative purposes, as the concept of "ownership" is inadequate to articulate Indigenous relationalities with land and more-than-human kin.

2 The hashtag #LandBack originated from a 2018 post by Blackfoot creator Arnell Tallfeathers. Journalist Riley Yesno defines #LandBack as follows: "Land Back is any action taken with the purpose of returning jurisdiction, authority, and resources to Indigenous people. This might include taking land back into Indigenous stewardship, restoring Indigenous people's legal rights to their land, or the active refusal to follow colonial laws on traditional and unceded territories." Since the term was coined in 2018, #LandBack has become a transnational movement fighting for Indigenous sovereignty over Indigenous homelands. Learn more about the history of the term here: https://newint.org/features/2022/10/24/land-back-decolonize-how

illustrating the land dispossession that is key to their very existence. Many universities' investment portfolios also include significant amounts of real estate, enabling these institutions to continue profiting from colonial land theft (Liu, 2023).

Additionally, research by Ahtone (Kiowa Tribe) and Lee (2020) shows that more than fifty American universities have benefited financially from Indigenous lands expropriated by the Morrill Act. Signed into law by President Lincoln in 1862, the Morrill Act "distribute[d] public domain lands to raise funds for fledgling colleges across the nation" (Ahtone & Lee, 2020, para. 3). The 10.7 million acres of land included in the Morrill Act were used not only for campuses, but were also monetized to provide seed funding for a number of institutions (ISSI, 2020). As Ahtone and Lee (2020, para. 10) write, "land-grant universities were built not just on Indigenous land, but with Indigenous land." Much of the Morrill Act lands were distributed with minimal compensation to Native nations; for more than 10 million acres taken from about 200 Tribes, the United States government paid only $400,000. No money was paid for over a quarter of the Morrill Act parcels, which were taken through treaties that were ultimately never ratified. Additionally, the Morrill Act required for its funds to be used in perpetuity, meaning that many universities continue to profit from the Morrill Act's settler colonial resource extractions. Funds from Morrill Act land grants were estimated to be about half a billion dollars in 2020 (Ahtone & Lee, 2020). Though Cal Poly Humboldt is not a land grant institution, its location on stolen Wiyot land indicates its settler colonial underpinnings.

Settler colonial universities' physical and financial foundations on stolen Indigenous lands are also reflected in their legacies of appropriating and claiming ownership of Indigenous knowledges and bodies, often while decentering and dehumanizing Indigenous peoples. As Māori scholar Linda Tuhiwai Smith writes, settler colonial research universities are "inextricably linked to European imperialism and colonialism" (2021, p. 1). Examples of colonial research abound in the archives, ranging from European travelers' tales that exoticized Indigenous peoples around the world to settler "scholar" Samuel Morton's racially coded skull measurements in *Crania Americana* (1839). Smith describes the effects of this compounded historical harm on Indigenous peoples in *Decolonizing Methodologies*:

> It appalls us that the West can desire, extract and claim ownership of our ways of knowing, our imagery, the things we create and produce, and then simultaneously reject the people who created and developed those ideas and seek to deny them further opportunities to be creators of their own culture and own nations. (2021, p. 1)

As a settler colonial institution, Cal Poly Humboldt has historically been complicit in this cycle of extraction from Indigenous peoples and knowledges. Simultaneously, due largely to the efforts of Indigenous students, staff, and faculty, the university has developed a plethora of Indigenous-facing programming over the past decades. Cal Poly Humboldt is located in Goudi'ni, which means "over in the woods" or "among the redwoods." This is the homeland of the Wiyot people, who continue to live in, tend to, and be in relationship with this area (Wiyot Tribe, n.d.). In addition to the Wiyot Tribe, the Humboldt area is also home(land) to "11 federally recognized and unrecognized tribes, rancherias and sovereign tribal governments," including the Yurok Tribe, Karuk

Tribe, and Blue Lake Rancheria (Department of Social Work, n.d.). Analysis of U.S. Census data by the Humboldt Area Foundation and Wild Rivers Community Foundation (2021) shows that Humboldt has "the most concentrated group of Native American tribes in the state of California" (para. 4). Cal Poly Humboldt is home to a significant number of Indigenous-facing academic programs, including the Indian Tribal & Educational Personnel Program (ITEPP), the Indian Natural Resources, Science and Engineering Program (INRSEP), the Rou Dalagurr Food Sovereignty Lab, the Goudi'ni Native American Arts Gallery, and the first Native American Studies B.A. program in the California State University system (Native American Studies, n.d.). These efforts have predominantly been led by Indigenous faculty, staff, and students, with varying levels of institutional support. In his 1989 report on *American Indians and Humboldt State*, Standing Rock Lakota scholar Vine Deloria Jr. wrote that "[t]he most impressive part of my visit to Humboldt State was to see how much the existing staff has accomplished with minimal to barely existing resources" (Recommendation 8). About thirty years later, Deloria's statement was exemplified in the creation of the Rou Dalagurr Food Sovereignty Lab, which was funded almost entirely by Native American Studies students, staff, faculty, and community members with minimal seed funding from the university (HSUNAS, 2021, 3:11).

Cal Poly Humboldt's Commitments to Decolonization

> What does it mean to be a 21st Century polytechnic university that is occupying dispossessed Indigenous lands and continues to benefit from Indigenous lands and resources? How do we at Cal Poly Humboldt engage with and foreground an ethical praxis of being in place? (Risling Baldy, Reed & Begay, 2023, p. 35)

Cal Poly Humboldt's planning documents reflect the university's engagement with this question and commitment to uplifting Indigenous communities and knowledges. In 2020, Humboldt State University (HSU) was in conversation about becoming the third polytechnic university in the California State University system. Published in 2021, HSU's Polytechnic Prospectus outlines steps needed for the university to become a polytechnic institution. In the prospectus, the move towards a polytechnic is framed as addressing this century's most pressing issues, with a particular focus on climate resilience and natural resource management (Humboldt State University, 2021). The prospectus makes numerous references to collaborating with Indigenous people and uplifting traditional ecological knowledges (TEK) as part of this effort. It highlights the university's "relationships with local tribes," arguing that these partnerships create "unparalleled opportunities for students to learn about the incorporation of TEK with Western scientific traditions for a fuller and more grounded education" (Humboldt State University, 2021). Mentions of TEK continue throughout the prospectus; in total, the document mentions "terms like 'tribe,' 'Indigenous,' 'Native American,' 'Indian,' and TEK" almost 200 times (Risling Baldy, Reed & Begay, 2023, p. 39).

Cal Poly Humboldt's 2021-2026 Strategic Plan reinforces these commitments, listing Decolonized Knowledge as one of the university's Core Values and Beliefs. "We value diverse systems of knowledge and ways of knowing," the document states, affirming a belief in "centering stories and knowledge that have been marginalized in oppressive social arrangements" (Humboldt State University, 2020, p. 9). The Strategic Plan also articulates a vision for the university that includes "partner[ing] with Indigenous communities to address the legacy of colonialism, and create space nurturing of traditional ecological knowledge (TEK), pedagogies, and curricula responsive to their identified needs" (Humboldt State University, 2020, p. 9). Tribal partnerships are also critical to Cal Poly Humboldt's climate resilience efforts. The university's 2021-22 Climate Action Plan includes a commitment to "Honoring the Land and its Original Inhabitants" in which the university commits to "draw [...] upon local Indigenous knowledge" to honor "life forms that have persisted against the colonial and extractive pursuits of settlers" (Cal Poly Humboldt, 2021, p. 1).

As demonstrated above, Cal Poly Humboldt has made many ambitious commitments to decolonization and partnerships with Indigenous peoples in its guiding documents. These commitments are aligned with the university's broader diversity and equity goals, and have played an important role in the university's polytechnic designation. However, work remains to be done for Cal Poly Humboldt to actualize these decolonial commitments. This includes supporting the Native American Studies department and other Indigenous-facing programming on campus, as Cutcha Risling Baldy (Hupa, Yurok and Karuk), Kaitlin Reed (Yurok, Hupa, and Oneida), and Kayla Begay (Hupa, Karuk, and Yurok) write (2023). The polytechnic funding that Cal Poly Humboldt received by promoting Indigenous knowledges must be allocated to Indigenous knowledge practitioners both within and outside the institution. This is a critical step towards repairing Cal Poly Humboldt's history of "profit[ing] off of Indigenous lands and knowledges—at the expense of Indigenous peoples" (Risling Baldy, Reed & Begay, 2023, p. 44).

In addition to meaningful support of anti-colonial knowledge systems, Cal Poly Humboldt must also apply this decolonial framework to its landholdings. True decolonial action goes beyond "decolonizing" one's bookshelf or curriculum and involves "the repatriation of Indigenous land and life" (Tuck & Yang, 2012, p. 1). As Risling Baldy, Reed, and Begay argue, "[t]here needs to be meaningful support from higher education institutions like Cal Poly Humboldt for Indigenous led movements like #LandBack" (2023, p. 45). My analysis of the Schatz Tree Farm outlines one such opportunity for Cal Poly Humboldt to support #LandBack efforts by returning a 385-acre forest that is currently under university ownership.

Schatz Tree Farm Overview

In the early 1970s, Los Angeles-based photographer Gordon Schatz and his wife decided to leave the city with their school-aged children as part of the back-to-the-land movement[3] While scanning a real estate catalog, Gordon was drawn to available tracts of land in Humboldt County. The area had been clear-cut by loggers in the 1950s, and real estate prices had dropped significantly after the cancellation of the proposed Butler Valley Dam

3 As outlined by Environment & Community alumnus Laura Hurwitz in 2017, the back-to-the-land movement is part of a larger pattern of settler colonial land dispossession. By asking the key question "back to whose land?," Hurwitz argues that settler homesteading is predicated on the theft of Indigenous lands.

project in the 1960s. With the help of his father L.W. Schatz, a CEO in the plastics industry, Gordon and his family purchased a 373-acre tract of land just north of Maple Creek, CA. They made their home there in the coming decades, becoming largely self-sufficient as they grew their own food and harvested timber. Gordon dedicated himself to managing the second-growth forest that had replaced the clear-cut, guided by a number of forestry textbooks his father passed down to him (Nichols, 2020). In 1987, the Schatz family decided to donate 385 acres of this property to Cal Poly Humboldt "to provide a demonstration tree farm operation for the benefit of the students and faculty of Cal Poly Humboldt and as an example for owners of small timberland parcels" (Cal Poly Humboldt, n.d.).

This narrative of the Schatz Tree Farm is presented throughout Cal Poly Humboldt's literature and documented history of the Tree Farm. However, this settler-focused history ignores the Whilkut people and other Indigenous peoples who had relationships with that land since time immemorial. In this paper, I will examine the ways in which settler colonialism is present in Cal Poly Humboldt's narratives around the Schatz Tree Farm. I will also attempt to answer Risling Baldy, Reed, and Begay's question of how "we at Cal Poly Humboldt [can] engage with and foreground an ethical praxis of being in place" (2023, p. 45) by reframing the university's relationship with the Tree Farm.

Methods

In order to interrogate settler narratives about the Schatz Tree Farm, textual analysis was conducted on several documents within Cal Poly Humboldt's Tree Farm archive. These include public-facing materials, such as the Tree Farm's website and its 2013 - 2016 annual reports. I also examined the history of the tree farm that's highlighted on its university webpage; published in 2020, this senior thesis by Forestry undergraduate Cole Nichols includes narrative oral history interviews with Tree Farm members and content analysis of the Tree Farm's historical documents. Some materials internal to the Forestry Department, such as the Tree Farm's land acknowledgement, were also included in this analysis. Created in November 2023, the land acknowledgement appears to be derived from the university's official land acknowledgement from the Office of the President and is accompanied by a fact sheet about the Whilkut people. This factsheet was created by the Califa Group, "a nonprofit library membership consortium" (2023), and draws from UC Berkeley anthropologist Alfred Kroeber's section on the Whilkut in his *Handbook of the Indians of California*.

Kroeber remains an influential figure in settler anthropology, as his ethnographies of Indigenous peoples shaped much of settler colonial discourse about California Indians. "According to many contemporary scholars," writes Cutcha Risling Baldy, "Alfred Kroeber was, and in many ways still is, the single greatest influence on the study of California Indian peoples" (2018, p. 74). However, as Risling Baldy points out, Kroeber's "adher[ence] to objectivity and positivism" (2018, p. 75) belied his own biases as a white settler man. Kroeber's documentation of California Indian culture focused primarily on material artifacts, failing to engage with the complexities of Indigenous spiritualities and philosophies (Risling Baldy, 2018). Notably, Kroeber remained silent on the issue of California Indian genocide, setting a precedent for the erasure of this genocide within the field of settler anthropology as a whole. Kroeber looms larger than life in settler studies of California Indians, and has been described as a "founder of modern American anthropology" (Stanlaw, 2022). As I will demonstrate in the following section, many of the forms of erasure in the Schatz Tree Farm's archives have their roots in Kroeber's work and worldviews.

My analysis of the Schatz Tree Farm archives is guided by frameworks of critical Indigenous studies, foregrounding Indigenous sovereignty within the U.S. settler colonial context. It builds on Indigenous-led scholarship regarding the relationships between universities and Native communities, including works by Robin Wall Kimmerer and Nicholas Reo. Theoretical foundations for this work include Saidiya Hartman's concept of silence in the archives, Patrick Wolfe's work on settler colonialism, and research by California Indian scholars Cutcha Risling Baldy and Kaitlin Reed on the history of both the Humboldt area and Cal Poly Humboldt itself.

Results

Patrick Wolfe (2006, p. 388) and Dean Saranillio (2015, p. 284) write that "settler colonialism destroys to replace;" "replacing one landscape for another, one people for another." In settler colonial narratives of the Schatz Tree Farm, it appears that one history has also been replaced in favor of another. A majority of Cal Poly Humboldt's archival documentation of the Tree Farm focuses on settler colonial histories and land management techniques, with Indigenous peoples relegated to the margins as part of what Wolfe calls the "settler-colonial logic of elimination" (2006, para. 1). In the few instances that Indigenous histories are included, they are conveyed in ways that erase the realities of genocide and land dispossession, locate Indigenous peoples in the past, and promote inaccurate understandings of Traditional Ecological Knowledge as simply "hunting and gathering."

The overwriting of Native histories with settler ones is evident in both the lack of Indigenous narratives and the erasure of genocide in the Tree Farm's published history. The tree farm's oral history document focuses primarily on settler histories and land management techniques, with less than one paragraph of the 50+ page publication pertaining to Indigenous peoples (Nichols, 2020). Similarly, a comprehensive review of the Tree Farm's website yields no mention of Indigenous peoples, though entire web pages are dedicated to the Tree Farm's history and goals and L.W. Schatz's contributions to the plastic industry (Cal Poly Humboldt, n.d.). The 2013 - 2016 Tree Farm annual reports, also posted on the website, mention Indigenous peoples only once and do not touch on the fact that the Tree Farm is Native land (Department of Forestry & Wildland Resources, 2013 - 2016). This singular mention is found in the 2015 annual report, which summarizes a visit to the farm by the cross-disciplinary Forestry Advisory Committee, which includes "representatives from [...] Native American Tribes" (Department of Forestry & Wildland Resources, p. 5). No further details are provided about which Tribes were represented or what their relationships with and goals for the Tree Farm were. Additionally, no tangible next steps or prospective avenues for Tribal collaboration are included in the report.

This erasure of Indigeneity was also written onto the Tree Farm's physical landscape. The Tree Farm and its accompanying research center had no land acknowledgements or other signage indicative of Native history and presence. Signage at the entrances to both the Tree Farm and the Research Center demonstrate the university's claim to the land, marking the area as Humboldt State University's L.W. Schatz Demonstration Tree Farm. Though seemingly innocuous ("Now they're taking issue with our name?"), Yurok and Karuk scholar Kerri J. Malloy argues that "[n]ames imposed by settler-colonists [are] part of taking possession of the land physically and metaphorically" (2023, p. 57). Critical geographer Reuben Rose-Redwood expands on this concept, writing that "[t]he naming of places is one of the primary ways in which the spatial imaginaries of colonialism have been entrenched within the spaces of everyday life in settler-colonial societies" (2016, p. 187). These "commemorative landscapes of settler colonialism" (Rose-Redwood, 2016, p. 187) serve to obscure the violent history of invasion and land theft that accompanied settlement, reaffirming the *terra nullius* myth and creating the illusion that the land has always belonged to settlers (Bigley, 2024).

The settler colonial nomenclature of the Tree Farm stands in stark contrast with Cal Poly Humboldt's campus, where places such as the Gutswurrak Student Activities Center, Goudi'ni Native American Arts Gallery, and the Native American Forum prominently display Indigenous names, signage, and artwork. Wiyot Plaza, the outdoor space between the Goudi'ni Gallery and the Native American Forum, was the most recent place on campus to be renamed ("'Wiyot Plaza' is New Name," 2022). These indigenized place names serve to "honor the Wiyot land where Cal Poly Humboldt sits" ("What's in a Rename?," 2022), "animating Indigenous history [and] reminding people of the ancestors of these lands" (Gray and Rück, 2019). Given the role of settler place names in Indigenous erasure, returning Indigenous nomenclature to the Tree Farm could serve as a powerful first step in " undermin[ing] the erasure of Indigenous people," as Malloy argues, "and recogniz[ing] the inherent indigenous character of the natural world" (2023, p. 62).

> [California Indians] transformed roots, berries, shoots, bones, shells, and feathers into medicines, meals, bows, and baskets and achieved an intimacy with nature unmatched by the modern-day wilderness guide, trained field botanist, or applied ecologist. (Anderson, 2005, p. 1)

In addition to its marginalization of California Indian history, genocide, and survivance, the Schatz Tree Farm archive also perpetuates the harmful stereotype that Indigenous people were merely hunters and gatherers. The single paragraph about the Whilkut people in the Tree Farm's oral history states that they "used the area for hunting and gathering" (Nichols, 2020, para. 11), vastly oversimplifying Indigenous relationalities to land and more-than-human kin. Often termed Traditional Ecological Knowledge (TEK), these systems of natural resource management (as they're referred to in the settler sciences) have been developed and refined through centuries of experimentation. M. Kat Anderson summarizes the pitfalls of this stereotype in *Tending the Wild*:

> The terms 'hunter-gatherer' and 'forager,' inaccurate anthropological labels assigned to most California Indian groups, connote a hand-to-mouth existence. They imply that California Indians dug tubers, plucked berries, and foraged for greens in a random fashion, never staying in any one place long enough to

leave lasting human imprints. (2005, p. 2)

This delegitimization of TEK plays into widespread settler narratives about California Indians. As early as 1869, settlers like John Muir marveled at California's landscape, stating that "[n]ature is the only gardener able to do work so fine" (Muir, 1911, Chapter 6). Muir couldn't conceive of California Indians as being able to manage such a complex landscape, viewing them instead as "passively gathering nature's 'wild' bounty" (Hogue, 2016, para. 3). In fact, what Muir failed to understand was that the "gardens" he gazed upon were not tended by Nature, but by the Indigenous Ahwanechee people of the Yosemite Valley. This was the case across California; as Jessica Hernandez (Maya Ch'orti' and Zapotec) writes, it was Indigenous people's knowledge of the land and ecological stewardship practices that "allowed them to steward [this] high biodiversity" that early settlers so admired (2022, p. 114). These stewardship practices were so successful that pre-invasion California "could support one of the highest densities of native people anywhere in the world" (Hogue, 2016, para. 11). Rather than simply being hunter-gatherers, the Whilkut and other California Indian peoples "were hunters, gatherers, harvesters, ecologists, botanists, biologists and much more" (Risling Baldy, 2018, p. 44).

Since "[t]he question of genocide is never far from discussions of settler colonialism" (Wolfe, 2006, p. 387), it follows that the silence in the Schatz Tree Farm's archives around settler colonialism also extends to the genocide that enabled that settlement. One notable example of this can be found in the Tree Farm's fact sheet on the Whilkut. Drawing on Kroeber's 1923 *Handbook* entry, the Whilkut population was estimated to be about 500 people in 1770, which dropped to only 50 in the 1910 Census (*Whilkut Tribe handout*, 2023). No further context or explanation is presented for this drastic decline. Similarly, Nichols states in the Tree Farm's oral history that "[s]harp decreases in [Indigenous] populations were the results of conflicts with white settlers" (2020, para. 11). Framing this genocide as simply a conflict or a drop in population belies the realities of genocide and dispossession that the Whilkut people faced, which Jack Norton (Hupa and Cherokee) details as follows:

> The Lassics, **Whilkuts**, Sinkyones, Mattoles, and nearly all the Wiyots, were annihilated as a people and a culture in the 'final solution to the Indian problem' by the representatives of a European white race who claimed superiority over all others. Those who were not so brutally attacked were forced to leave their homes, their ancestors' graves and their cosmographic realms that defined and held them together as a people. They were sold into slavery or marched hundreds of miles to reserves that were the least desirable lands in the eyes of the whites during those tragic days of removal. (1979, p. X, emphasis added)

These dual erasures of Indigenous presence and settler colonial genocide are part of a larger discursive trend in ethnohistorical research pertaining to California Indian people. This is exemplified by Kroeber's entry on the Whilkut in his *Handbook*, in which he dismisses the Whilkuts' presence on the land by stating that they "are practically unknown" (1925, p. 141). This assertion reinforces a settler colonial perspective when in fact, according to Kroeber's own writings (1925), the Whilkut had historically engaged in reciprocal relationships with neighboring tribal

Nations, non-human relatives, and the land itself. The passive language used in the Tree Farm's oral history to refer to the genocide also reflects similar perspectives shared by Kroeber. In his *Handbook*, Kroeber states that "the Whilkut suffered heavily in the same struggles with the whites which caused the Chilula to melt away" (1925, p. 141). Kroeber's use of the passive voice to refer to the Whilkut and his obfuscation of the genocide as "struggles"render the settlers blameless while making it seem like the Whilkut "melt[ed] away" of their own accord, leaving behind an uninhabited *terra nullius* that was ripe for settlement.

In addition to negating the Whilkut people's longstanding history on this land, the Schatz Tree Farm archives erase Whilkut survivance (Vizenor, 2008) by referring to them largely in the past tense. Both the Tree Farm's land acknowledgement handout and oral history are ripe with language that places Whilkut people in the past. For example, the land acknowledgement handout states that Whilkut people "were until recent years one of the dominant Athapaskan tribes of Humboldt County" (*Whilkut Tribe handout*, 2023, para. 3). It details where their villages "were located" (para. 8), how they "built their houses" (para. 10), and how "Whilkut women made baskets much like those of other early northwest California women" (para. 19). By temporally locating the Whilkut people in the past, these documents "perpetuate the myth that Indigenous Peoples [...] no longer exist as distinct cultures" (Indigenous Citation Guide, 2024).

As with other forms of settler colonial erasure, the Tree Farm's archives reflect a larger discursive trend in settler narratives about Indigenous peoples. A 2015 study by Sarah Shear shows that "[a] staggering 87 percent of references to American Indians in all 50 states' academic standards portray them in a pre-1900 context" (Landry, 2018). This, Shear argues, "direct[s] students to see Indigenous Peoples as a long since forgotten episode in the country's development," furthering "a Euro-American narrative that reinstitutes the marginalization of Indigenous cultures and knowledge" (Shear et al., 2015, p. 90). It also serves to minimize the continued survivance of Indigenous peoples; as Cutcha Risling Baldy and Kayla Begay write, "references only to our genocide, our victimization at the hands of egregious laws and depraved settlers, frame our people as passive victims who perished at the hands of settlers, thereby erasing our continued presence on this land" (2019, p. 42).

Despite enduring decades of genocidal violence, descendants of the Whilkut people continue to live within and be in relationship with their ancestral homelands. As part of an 1864 treaty with the United States government, "the Whilkuts [...] joined the Hupas on the newly established Hoopa Valley Reservation" (Rohde, Forthcoming, "Whilkut Territory"). Whilkut descendants continue to be enrolled in both the Hoopa Valley Tribe ("Hoopa Valley Tribe," 2022) and Blue Lake Rancheria (J. Rohde, personal communication, July 16, 2024). However, neither the Schatz Tree Farm's land acknowledgement nor any of its other documents make this connection between the Whilkut and current Native nations. Doing so, while most likely unintentional, serves to overlook the responsibilities the university and the Tree Farm have to these nations.

Conclusion

Though the Schatz Tree Farm has no public-facing land acknowledgement, its online archive on Google Drive, which is accessible only to select people primarily within the Forestry department, contains a land acknowledgement accompanied by a fact sheet on the Whilkut people. Modeled after the university's standard land acknowledgement, the document acknowledges that the Tree Farm is located on unceded Whilkut lands and affirms the Tree Farm's commitments to partnering with Indigenous peoples. These reflect the university-wide commitments and goals outlined in the Strategic Plan, Polytechnic Prospectus, and other guiding documents. The land acknowledgement states the following:

> We acknowledge that the L.W. Schatz Demonstration Tree Farm is located on the unceded lands of the Whilkut people, where they have resided from time immemorial. We encourage all to gain a **deeper understanding of their history and culture**. As an expression of our gratitude we are genuinely committed to **developing trusting, reciprocal, and long lasting partnerships** with the Whilkut people as well as all of our neighboring tribes. (*Whilkut Tribe handout*, 2023, para. 1, emphasis added)

While this land acknowledgement encourages readers to gain a deeper understanding of Whilkut land and culture, no other documents in the Tree Farm's archive provide information to further this understanding. Instead of providing a comprehensive and accurate history of this university-owned parcel, this statement shifts the responsibility of learning about the Whilkut people onto the individual reader. Additionally, the commitments expressed in this land acknowledgement have not historically been carried out. Despite its stated commitment to "developing trusting, reciprocal, and long lasting partnerships with the Whilkut people as well as all of our neighboring tribes" (*Whilkut Tribe handout*, 2023, para. 1), the Schatz Tree Farm has not partnered with any Native nations, Indigenous peoples, or Indigenous-facing entities on campus (E. Kelly, personal communication, October 8, 2024).

Though land acknowledgements are often an important first step towards allyship, in order to be meaningful, they must be accompanied by action that is supportive of Indigenous sovereignty. The Native Governance Center describes the importance of acting on land acknowledgements as follows:

> It's easy for land acknowledgments to become yet another form of optical allyship. They often lack a call to action and next steps. Without these components, land acknowledgments are just empty words. They become an excuse for folks to feel good and move on with their lives without actually contributing anything to the community. As President Robert Larsen of the Lower Sioux Indian Community puts it, "An apology or an acknowledgment is one thing, but *what are you going to do next?"* (2021, para. 6)

After giving a land acknowledgement, Dr. Cutcha Risling Baldy argues,

> Now it's time to think about what that actually means for you or your institution. What are the concrete actions you're gonna take? What are the ways you're gonna assist Indigenous peoples in uplifting and upholding their sovereignty and self-determination? (2023, para. 7)

As Risling Baldy et al. argue, returning land to Native nations can be an important step towards repairing extractive settler-colonial relationships and putting land acknowledgements into action (2024). Given the critical role of land acquisition in the colonization of Turtle Island (Dunbar-Ortiz, 2023), returning land must be an essential part of

Cal Poly Humboldt's commitments to decolonization. As Diné journalist Kalen Goodluck argues, "[f]or Native peoples, land is not just acreage, nor simply the foundation of tribal sovereignty under federal law. The land is also inextricably intertwined with personal and family histories, collective tribal identity, and cultural survival" (2023, para. 16). By returning land to Indigenous ownership and care, settler landowners like Cal Poly Humboldt can support Native nations' sovereignty over their ancestral homelands.

Though there have been numerous examples of land return in recent years (over 100 instances in the past two decades! (Goodluck, 2023)), returning the Schatz Tree Farm presents somewhat of a unique scenario given its role as a university-owned demonstration forest. However, a similar parcel of forestry land was returned by the University of Minnesota to the Fond du Lac Band of Lake Superior Chippewa Tribe earlier this year. Located within Fond du Lac ancestral territory, the 3,400-acre Cloquet Forestry Center "has been the primary research and education forest for the University of Minnesota" for decades, per an FAQ document issued by the university ("Frequently asked questions," 2024, para. 1). Similarly to the Tree Farm, the Cloquet Forestry Center was also clear-cut by logging companies prior to its sale to the university ("Frequently asked questions," 2024). Much of the research conducted at the Forestry Center parallels that of the Tree Farm, including silviculture, wildlife, ecology, and hydrology research. This research will likely continue after the land has been returned, as next steps in the land transfer include "shared commitments and collaboration on forestry and natural resources education and research" ("Frequently asked questions," 2024, para. 9). The University of Minnesota's pursuit of legislative avenues to return this land in 2025 may provide an actionable blueprint for Cal Poly Humboldt to return the Schatz Tree Farm.

We don't have to look as far as Minnesota to find examples of universities returning land to tribes; in fact, California State University, Chico (Chico State) has returned land to Native nations on two separate occasions. In September 2022, Chico State became the first university in California to return land to a tribe by transferring the 93-acre Butte Creek Ecological Preserve to the Mechoopda Tribe, on whose ancestral land the preserve is located. This land was transferred at no cost to the Tribe, and a Memorandum of Understanding (MOU) between Chico State and the Tribe allows faculty and students to access the property for research and educational purposes (Staples, 2022). One of the benefits of this land transfer is that it will allow the Mechoopda to apply their place-specific land management techniques to help restore the Preserve after a recent fire. Both the university and other partners in the land transfer recognized the Mechoopda Tribe's unique qualifications to manage this land with traditional stewardship techniques. As John Donnelly, executive director of the California Wildlife Conservation Board, stated:

> The Mechoopda, with centuries of expertise of our local ecosystems and watersheds, are uniquely qualified to improve and manage the Butte Creek Ecological Preserve. The Tribe's sophisticated restorative land management practices will help create a healthy ecosystem that is resilient, fire-adaptive, and beneficial to the entire community. (2022, para. 10)

One year after the return of Butte Creek Canyon, Chico State again gave #LandBack; this time, they returned the 73-acre Eagle Lake Field Station to the Susanville Indian Rancheria. The land return was made possible by the 2023 Recreation & Public Purposes Tribal Parity Act, which "allows the Bureau of Land Management (BLM) to approve the sale and leasing of land to Tribal governments for significantly below market value, as long it is used for recreational and other public purposes" (Staples, 2023).

As the land returns by Chico State and the University of Minnesota show, settler universities can begin to repair their legacy of colonial harm by returning Indigenous lands to Indigenous peoples. The legislative pathways utilized by both universities provide templates for Cal Poly Humboldt to engage in land return. By returning the Schatz Tree Farm, Cal Poly Humboldt can act on its Strategic Plan's goals of partnering with Indigenous communities and decolonizing the university's knowledge systems. Only by engaging in true decolonial action through land return can Cal Poly Humboldt begin to "engage with and foreground an ethical praxis of being in place" (Risling Baldy, Reed & Begay, 2023, p. 35).

As a settler institution with stated commitments to decolonization, Cal Poly Humboldt stands at an interesting crossroads as it transitions to a polytechnic university. While its guiding documents affirm the university's commitment to partnering with Indigenous communities, this commitment is not reflected in the university's history and narratives of one of its properties, the Schatz Tree Farm. Cal Poly Humboldt's archive on the Tree Farm presents a predominantly settler-colonial perspective of this place, thereby erasing Indigenous history, genocide, and survivance on and around the Tree Farm. As Cal Poly Humboldt strives to actualize the decolonial commitments made in its guiding documents, it is responsible for rebuilding trust and relationships with Native nations by returning stolen lands. I therefore urge Cal Poly Humboldt to follow the examples of the University of Minnesota and Chico State University by returning the Schatz Tree Farm to Native nations.

References

Ahtone, T. & Lee, R. (2020, March 30). *Land-grab universities*. High Country News. https://www.hcn.org/issues/52-4/indigenous-affairs-education-land-grab-universities/

Anderson, K. (2005). *Tending the wild: Native American knowledge and the management of California's natural resources*. Univ of California Press.

Bigley, M. (2024, July 5). *The power of reclaiming Indigenous place names*. Save The Redwoods League. https://www.savetheredwoods.org/blog/the-power-of-reclaiming-indigenous-place-names/

Cal Poly Humboldt. (n.d.). *L.W. Schatz Demonstration Tree Farm*. Cal Poly Humboldt. https://www.humboldt.edu/tree-farm

Cal Poly Humboldt. (2021). *Climate Action Plan 2.0 Strategies to Achieve Carbon Neutrality and Climate Resilience*. Cal Poly Humboldt. https://www.humboldt.edu/sites/default/files/sustainable-campus/2024-07/cap20final4.pdf

Califa Group. (2023). *About Us*. Califa. https://califa.org/about

Deloria, Vine Jr. (1989). *Consultant Report: American Indians and Humboldt State*. Cal Poly Humboldt. https://scholarworks.calstate.edu/concern/publications/9w0325368

Department of Forestry & Wildland Resources.

(2013). *L.W. Schatz Demonstration Tree Farm 2013 Annual Report*. Humboldt State University. https://www.humboldt.edu/sites/default/files/tree-farm/2024-08/112013annual-report.pdf

Department of Forestry & Wildland Resources. (2014). *L.W. Schatz Demonstration Tree Farm 2014 Annual Report*. Humboldt State University. https://www.humboldt.edu/sites/default/files/tree-farm/2024-08/112014annual-report.pdf

Department of Forestry & Wildland Resources. (2015). *L.W. Schatz Demonstration Tree Farm 2015 Annual Report*. Humboldt State University. https://www.humboldt.edu/sites/default/files/tree-farm/2024-08/112015annual-report.pdf

Department of Forestry & Wildland Resources. (2016). *L.W. Schatz Demonstration Tree Farm 2016 Annual Report*. Humboldt State University. https://www.humboldt.edu/sites/default/files/tree-farm/2024-08/112016annual-report.pdf

Department of Social Work. (n.d.). *Geographical Context*. Cal Poly Humboldt. https://socialwork.humboldt.edu/content/geographical-context

Dunbar-Ortiz, R. (2023). *An indigenous peoples' history of the United States. Beacon Press.*

Frequently asked questions on intentions for Cloquet Forestry Center. (2024, May 24). University of Minnesota. https://drive.google.com/file/d/1-f2krp3S1iDWrwPX4zU9bX-v4SXobX9ud/view

Goodluck, K. (2023, September 11). *The Land Back Movement Unravels Manifest Destiny*. Sierra: The Magazine of the Sierra Club. https://www.sierraclub.org/sierra/2023-3-fall/feature/land-back-movement-unravels-manifest-destiny

Gray, C. & Rück, D. (2019, October 8). *Reclaiming Indigenous Place Names*. Yellowhead Institute. https://yellowheadinstitute.org/2019/10/08/reclaiming-indigenous-place-names/

Hartman, S. (2008). Venus in Two Acts. *Small Axe, Number 26 (12*, 2), 1-14. https://warwick.ac.uk/fac/arts/history/research/centres/blackstudies/venus_in_two_acts.pdf

Hernandez, J. (2022). *Fresh banana leaves: Healing Indigenous landscapes through Indigenous science*. North Atlantic Books.

Hogue, L. (2016, September 16). *What John Muir Missed: The Uniqueness of California Indians*. PBS SoCal. https://www.pbssocal.org/shows/tending-the-wild/what-john-muir-missed-the-uniqueness-of-california-indians

Hoopa Valley Tribe. (2022). K'ima:w Medical Center. https://www.kimaw.org/hvt

HSUNAS. (2021, February 27). *Telling the Story of the NAS Food Sovereignty Lab & Cultural Workspace* [Video]. YouTube. https://www.youtube.com/watch?v=prMerJPYXgc&list=PLQjn8Pj_cTF-anPTGse-nspTW760GgpJ6&index=12

Humboldt Area Foundation & Wild Rivers Community Foundation. (2021). Humboldt Area Foundation + Wild Rivers Community Foundation Service Area: A Regional Context. *COVID-19 Regional Response Fund Report Appendix 1: Regional Context*. https://hafoundation.org/Portals/0/Uploads/Documents/Publications/COVID_Report/Appendix_1V2.pdf

Humboldt State University. (2020). *Future*

Forward: Strategic Plan 2021-2026. Cal Poly Humboldt. https://strategicplan.humboldt.edu/sites/default/files/strategicplan_2020_final.pdf

Humboldt State University. (2021, September 1). *Polytechnic Prospectus*. Cal Poly Humboldt. https://www.humboldt.edu/sites/default/files/2023-01/hsupolytechnicprospectus.pdf

Hurwitz, L. S. (2017, December 21). *Settler colonialism and white settler responsibility in the Karuk, Konomihu, Shasta, and New River Shasta Homelands: a white unsettling manifesto*. Cal Poly Humboldt. https://digitalcommons.humboldt.edu/etd/106/

Indigenous Citation Guide. (2024, November 8). University of British Columbia. https://guides.library.ubc.ca/c.php?g=721560&p=5161438#

ISSI. (2020, September 25). *UC Land Grab: A Legacy of Profit from Indigenous Land - Part 1* [Video]. YouTube. https://www.youtube.com/watch?v=6z0QbDz3KDE

Kroeber, A. L. (1925). *Handbook of the Indians of California* (Vol. 78). US Government Printing Office.

Landry, A. (2018, September 13). *'All Indians Are Dead?' At Least That's What Most Schools Teach Children*. ICT. https://ictnews.org/archive/all-indians-are-dead-at-least-thats-what-most-schools-teach-children

Liu, C. (2023, March 8). *How universities became giant piggy banks for hedge-fund billionaires*. Business Insider. https://www.businessinsider.com/universities-colleges-turning-into-real-estate-hedge-funds-higher-education-2023-3

Malloy, K. J. (2023, October 13). Naming to Erase, Naming to Restore. In Gensburger, S., & Wüstenberg, J. (Eds.). *De-commemoration: Removing statues and renaming places* (Vol. 12). Berghahn Books.

Morton, S. G., & Combe, G. (1839). *Crania Americana; or, a comparative view of the skulls of various aboriginal nations of North and South America: to which is prefixed an essay on the varieties of the human species*. Philadelphia: J. Dobson; London: Simpkin, Marshall.

Muir, J. (1911). *My First Summer in the Sierra*. The Riverside Press Cambridge.

Native American Studies. (n.d.). *Home*. Cal Poly Humboldt. https://nasp.humboldt.edu/nas-welcome

Native Governance Center. (2021, October 11). *Beyond Land Acknowledgment: A guide for building meaningful relationships with Native Nations*. Native Governance Center. https://nativegov.org/news/beyond-land-acknowledgment-guide/

Nichols, C. (2020). *Integrating Legacy Stewardship Knowledge in Managing for the Future: An Oral History of the L. W. Schatz Demonstration Tree Farm*. Humboldt State University. https://docs.google.com/document/d/e/2PACX-1vQS975vfQeM-v68gI6lq6382AW3XhFpOVe-Da6eK4YvOVSdRmk4On03qJVhVBqO-J0AYuetuGVyWgz0RA0/pub

Norton, J. (1979). *Genocide in Northwestern California: When Our worlds Cried*. The Indian Historian Press.

Office of the President. (2023). *Land Acknowledgement*. Cal Poly Humboldt. https://president.humboldt.edu/land-acknowledgement

Risling Baldy, C. (2018). *We are dancing for*

you: native feminisms and the revitalization of women's coming-of-age ceremonies. University of Washington Press.

Risling Baldy, C. & Begay, K. (2019). Xo'ch Na:nahsde'tł'-te: Survivance, Resilience, and Unbroken Traditions in Northwest California. In Lara-Cooper, K. & Lara Sr., Walter J. (Eds.). *Ka'm-t'em: A Journey Toward Healing*. Great Oak Press.

Risling Baldy, C., Reed, K. P., & Begay, K. (2023). Polytech to PolyTEK: Traditional Ecological Knowledge, Indigenous Science, and the Future Forward Polytechnic University. *Humboldt Journal of Social Relations*, 1(45), 34-51.

Risling Baldy, C., Tully, C., & Rojas, K. M. (2024). California #LandBack Special Report. Save California Salmon. https://www.californiasalmon.org/_files/ugd/d97ff6_872841991d6f4030b2391e466e61c908.pdf

Rohde, Jerry. (Forthcoming). *Northern Humboldt Indians*. The Press at Cal Poly Humboldt.

Rose-Redwood, R. (2016). " Reclaim, Rename, Reoccupy": Decolonizing Place and the Reclaiming of PKOLS. ACME: *An International E-Journal for Critical Geographies, 15*(1).

Saranillio, D. I. (2015). Settler colonialism. *Native studies keywords*, 284-300.

Shear, S. B., Knowles, R. T., Soden, G. J., & Castro, A. J. (2015). Manifesting destiny: Re/presentations of indigenous peoples in K–12 US history standards. *Theory & Research in Social Education, 43*(1), 68-101.

Smith, L. T. (2021). *Decolonizing methodologies: Research and indigenous peoples*. Bloomsbury Publishing.

Solnit, R. (2021, March 2). *John Muir in Native America*. Sierra: The Magazine of the Sierra Club. https://www.sierraclub.org/sierra/2021-2-march-april/feature/john-muir-native-america

Stanlaw, J. (2022). Alfred Kroeber and the Development of Linguistic Anthropology: A Brief Reassessment| Bérose. *Un article du dossier documentaire: Kroeber, Alfred Louis (1876–1960)*.

Staples, A. (2022, September 23). *Ancestral Land in Butte Creek Canyon Returned to the Mechoopda Tribe. Chico State Today.* https://today.csuchico.edu/bcep-transfered-to-mechoopda-tribe/

Staples, A. (2023, November 13). *Chico State Enterprises Returns Eagle Lake Field Station Land to Susanville Indian Rancheria*. Chico State Today. https://today.csuchico.edu/chico-state-returns-eagle-lake-field-station/

Tervalon, M., & Murray-Garcia, J. (1998). Cultural humility versus cultural competence: A critical distinction in defining physician training outcomes in multicultural education. *Journal of health care for the poor and underserved, 9*(2), 117-125.

Tuck, E. & Yang, K. W. (2012). Decolonization is not a metaphor. *Decolonization: Indigeneity, Education & Society, 1*(1), 1-40. https://clas.osu.edu/sites/clas.osu.edu/files/Tuck%20and%20Yang%202012%20Decolonization%20is%20not%20a%20metaphor.pdf

Veltman, C. (2023, March 15). *So you began your event with an Indigenous land acknowledgment. Now what?*. NPR. https://www.npr.org/2023/03/15/1160204144/indigenous-land-acknowledgments

Vizenor, G. (Ed.). (2008). *Survivance: Narratives of native presence*. U of Nebraska Press.

What's in a Rename? A Building that Honors the Wiyot People and Land. (2022, April 21). Humboldt Now. https://now.humboldt.edu/news/whats-rename-building-honors-wiyot-people-and-land

Whilkut Tribe handout. (2023, November 3). Records & Documents. https://drive.google.com/file/d/1ZozthfZmwE_U0dUUgFrOx-h5LOmTi-BAC/view?usp=sharing

Wiyot Tribe. (n.d.). Tribe History. *Wiyot Tribe - Loleta, California*. https://wiyot.us/

'Wiyot Plaza' is New Name for Campus Space. (2022, November 7). Humboldt Now. https://now.humboldt.edu/news/wiyot-plaza-new-name-campus-space

Wolfe, P. (2006). Settler Colonialism and the Elimination of the Native. *Journal of genocide research, 8*(4), 387-409.

IdeaFest
Journal

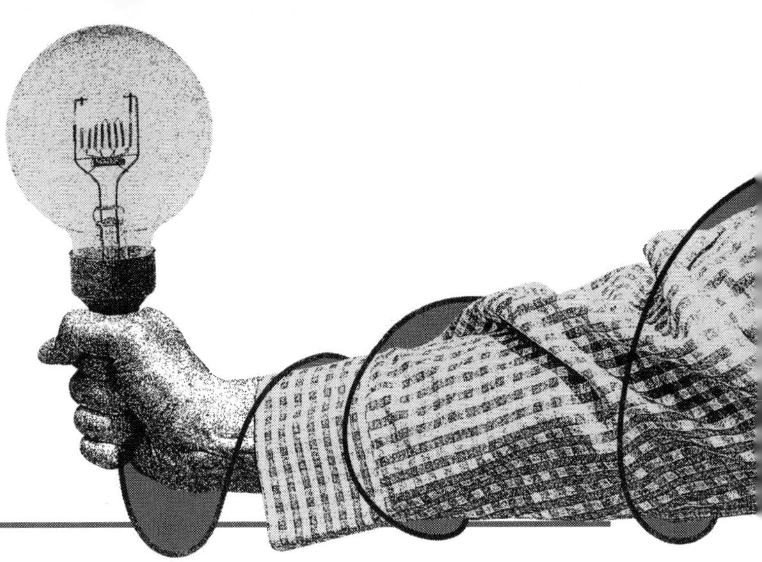

Refining a Flow Injection Analysis Method for Trace Concentrations of Iron in Seawater

Brooke I. Stafford and Elijah J. Vestal, Department of Geology
Liam W. Hodgson, Department of Physics & Astronomy
Noah S. Schuhmann, Department of Biological Sciences
Claire P. Till, Department of Chemistry & Biochemistry
Cal Poly Humboldt

Iron is an essential micronutrient required for the growth of phytoplankton, the base of the oceanic food chain. Iron exists in the ocean in extremely low concentrations, limiting primary productivity in ~⅓ of the surface ocean, and can be affected by physical processes such as upwelling of water masses, and by biological and chemical processes within the water. In order to further our understanding of iron availability in the ocean, here we refine and improve our understanding of an analytical method to measure nanomolar iron in seawater. The method used is an adaptation of Lohan et al. (2006), a flow injection analysis method with catalytic spectrophotometric detection. Here, we quantify the blank and its precision; we investigate the effect of different concentrations, pHs, and reagent qualities of the ammonium acetate buffers on the blank and baseline signal; and we investigate the limitation imposed by the data acquisition system (DAQ), which converts analog signal into digital readings. We found that our blank has an average concentration of 0.53 ± 0.07 nM; that the pH of the rinse solution does not have a significant impact on the peak shape, but that its concentration does; that reagent quality of the buffer and rinse solutions does significantly affect our blank and baseline; and that the two DAQ models tested have a lower limit on their detection range of ~0.03 nM.

Introduction

Background

Iron is a vital trace metal for marine life, being a limiting nutrient in about 1/3 of the surface ocean (Moore et al., 2004). Iron typically exists in nanomolar concentrations as oxidized iron (Fe^{3+}) in marine settings due to the availability of dissolved O_2. Oxidized iron is insoluble in seawater, so the vast majority of iron input to the ocean, via rivers or dust for example, sinks and is removed from the water column where it is unavailable to surface dwelling phytoplankton.

Having a better understanding of the distribution of iron in seawater provides insight into how iron and marine ecosystems interact.

However, the analysis of nanomolar concentrations of iron is a challenge: not only are nanomolar concentrations of iron inherently difficult to measure, but the seawater samples also contain high concentrations of salts, further complicating the analytical procedure. Many methods exist to overcome these challenges, but they can be finicky. Here we work to refine and better understand one of these methods for accurately measuring nanomolar quantities of iron in seawater, in order to contribute to the further understanding of the marine ecosystem.

General Method

We utilize the flow injection analysis (FIA) method of Lohan et al. (2006), with modifications described in Biller et al. (2013) to measure nanomolar iron in seawater. This method utilizes a six-channel peristaltic pump rotating at 4.5 RPM to concentrate iron on atoyopearl chelate-650 resin column before elution into a reaction stream of different reagents (Figure 1).

The sample is loaded onto the column at a pH of 2. At this pH, upwards of 93% of Fe3+ will bind to the column (Biller et al., 2013). In order to keep

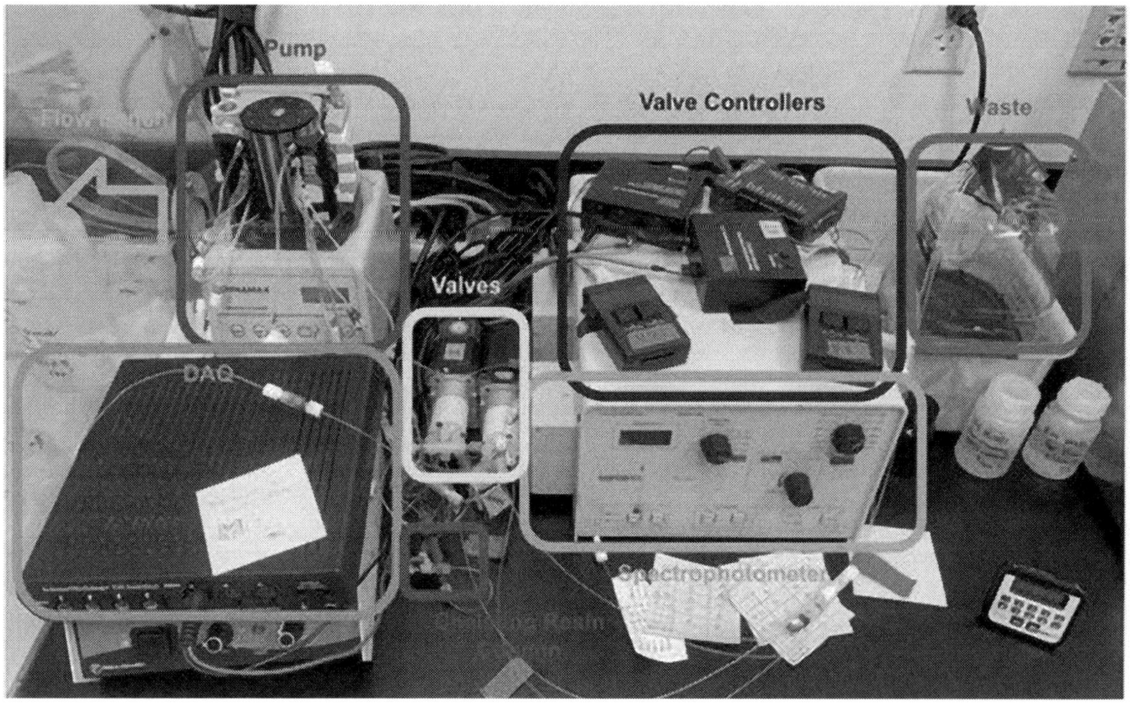

Figure 1: The method setup. The flow bench (left) is a trace metal clean space that contains the sample and reagent bottles. Plastic tubes containing the sample and reagents emerge from the back of the flow bench and reach a pump (red), which pushes the solutions through mixing coils, valves and chelating resin columns until the reaction stream reaches the spectrophotometer (orange). The DAQ (green) takes the analog signal from the back of the spectrophotometer and converts it into a digital signal. The valve controllers (blue) set the valves (yellow), which determine whether the system is rinsing, loading, or eluting. Ultimately, everything ends up in the waste (pink).

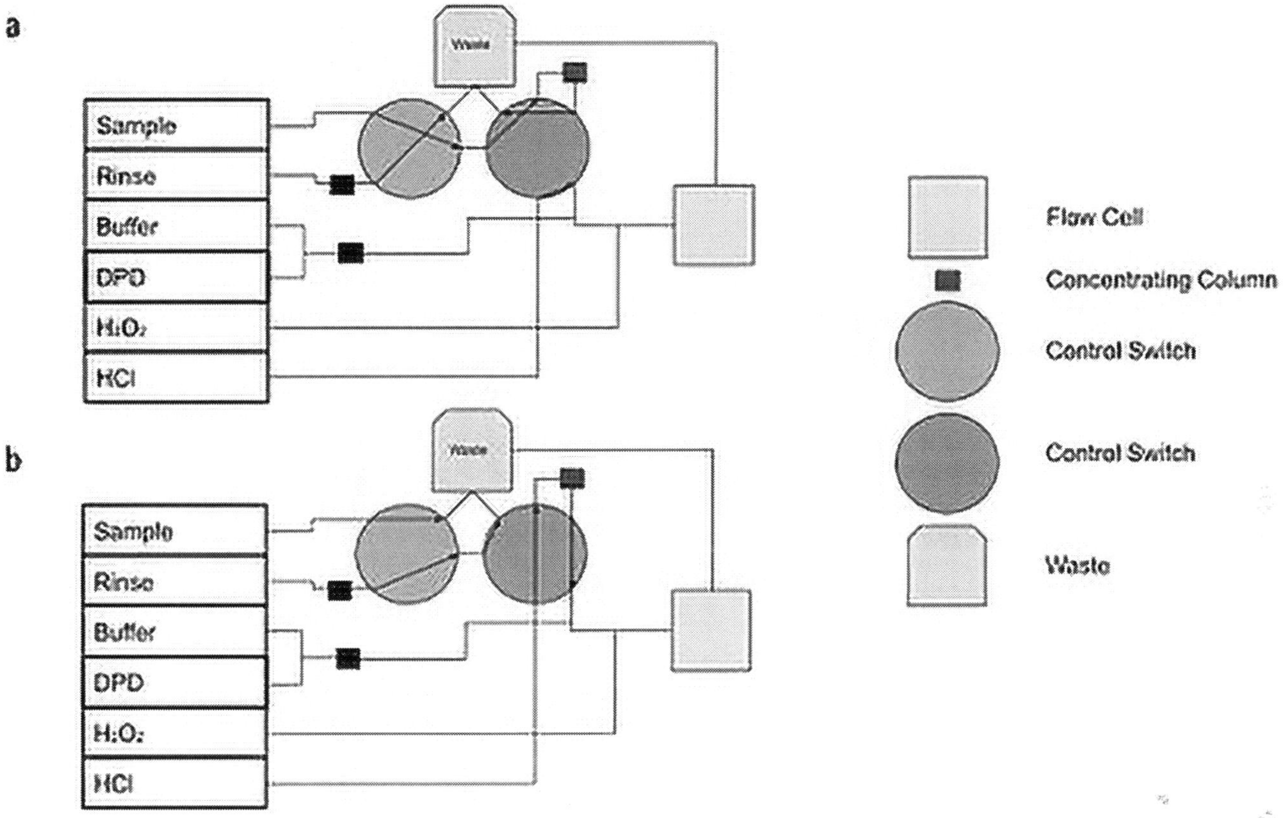

Figure 2: Schematic of the preconcentration flow injection analysis setup for a) loading and b) eluting a sample.

the iron in this oxidized form, hydrogen peroxide (H2O2, final concentration 10 μM) is added to each sample before loading it on the column.

The process for preconcentrating each sample, standard or blank is as follows. After the previous sample is eluted with 1-1.5 M optima HCl, the pH of the column is too low to bind iron. A 3.25 M ammonium acetate rinse solution at a pH of ~3-3.5 is sent through to condition the column to a higher pH so that iron will bind. Then, the sample is loaded onto the column in the load phase (Figure 2). Depending on the specified iron concentration, samples can be loaded for anywhere between 30 seconds to 4 minutes. After a sample is loaded, the rinse solution is sent through again, this time to remove accumulated salts from the column. Then, the concentrated sample is eluted with the optima HCl into the reaction stream (see below). This process is repeated for each sample, standard and blank.

The reaction stream contains DPD (N,N-dimethyl-p-phenylenediamine), H2O2 and a 3.5M ammonium acetate buffer solution (pH ~9), which mix with HCl that has either just gone across the column (during the elute phase), or (in the condition, load, and rinse phase) bypasses the column (Figure 2). This reaction stream passes through mixing coils where it darkens depending on the iron concentration: as the Fe3+ oxidizes the individual DPD molecules, they are turned pink as the Fe3+ is reduced to Fe2+. The addition of H2O2 to the system makes the Fe2+ oxidize back to Fe3+ so that more DPD molecules can be oxidized by that individual iron atom. In this way, the signal of the small iron concentration is amplified.

The buffer (ammonium acetate) in the reaction stream is at a pH of 9. The combination of the HCl (pH <0), the buffer (pH ~9), and the H2O2, set this reaction stream at a pH of ~5.7, which is ideal for the reaction that occurs between the DPD and Fe3+. After the DPD is oxidized, it is sent through the spectrophotometer to be measured for absorbance. The spectrophotometer outputs an analog signal in the form of a voltage difference across two screws. This signal corresponds with absorbance. The data acquisition system (DAQ) is wired to both screws so it can read the voltage difference and send it to the computer as a digital signal. The absorbance versus time is recorded, and we are able to process it in excel. The peak heights correlate with iron concentration.

To quantify the concentration in seawater samples, a sample with a relatively low iron concentration is spiked with increasing amounts of iron. Each of these is measured, and a standard curve is constructed based on those measurements. This method, called standard additions, is used because of the salts present in the seawater samples. The salts can affect the analysis of the samples, so we want the standards to be impacted in the same way. The seawater used to create the standards contains some iron, so the y-intercept is not representative of a blank. For this reason, we use only the slope of the standard curve to analyze peak heights.

Optimizing the Method

The goal of our recent work has been to investigate uncertainties of the method in several ways: quantifying the blank and detection limit at three sample load times, assessing the impact of changing the pH and concentration of the ammonium acetate (NH4Ac) rinse, comparing HPLC, Optima, and quartz-distilled reagents for the buffers, and quantifying signal limitations of the data acquisition system.

Methods

Blank Measurements

The goal of obtaining blank measurements was to be able to assess modifications to the system. Blanks were thus measured by analyzing MilliQ in place of a sample, and otherwise sticking exactly to the method described in section 1.2. We measured for a load time of one, two, and four minutes, 4-5 times each. The values of our measured absorbance were compared to known standard concentrations to find the concentrations of the blanks.

We assume that MilliQ has insignificant amounts of iron, and thus the peak size of the blank corresponds with the signal from the rest of the system, potentially from minor contaminants in our reagents, the preconcentration procedure, or detection limits from the DAQ.

Buffer Grade

Historically our lab group—and indeed, the lab groups of Lohan et al. (2006) and Biller et al. (2013)—have used quartz distilled acetic acid bubbled with ammonia to make trace metal clean ammonium acetate crystals. These crystals were then dissolved to make a saturated solution of ammonium acetate, which was then diluted as described below to make the rinse and buffer solutions for our FIA method. However, this process to make the quartz distilled clean ammonium acetate is laborious and involves hot fuming solutions. Our lab group has a finite supply of crystals made this way. Here, we investigate alternatives to quartz-distilled crystals that can be cleanly purchased.

We evaluate the effect of the rinse and buffer reagent quality on blank absorbance values by purchasing separate HPLC grade and Optima grade ammonium acetate and making saturated solutions from these crystals. Optima glacial acetic acid and optima ammonium hydroxide were also purchased and prepared at a later date.

For all reagent qualities (quartz-distilled, optima, and HPLC), solutions of ammonium acetate were made and diluted with MilliQ to 3.25 M for the rinse and 3.5 M for the buffer. The pH of the rinse was adjusted to 3-3.5 and the buffer to a pH of ~9 using optima HCl or optima ammonium hydroxide, as needed. Then, the pH of the reaction stream mixture was tested by mixing eluent HCl and buffer in a 1:3 ratio (the ratio they are mixed in the reaction stream) and ensuring it was pH 5.6-5.8. The pH and/or concentration of the rinse, buffer, and HCl solutions were adjusted until the pH of each solution and ratio were correct. (This was also done with mixing the rinse and buffer solutions in a 1:3 ratio, because there is a small amount of time in the reaction stream, just after rinsing the column, when rinse solution is in the reaction stream instead of HCl.)

MilliQ blanks and iron-spiked standards were used rather than samples, letting us quantify differences among the four buffers. We acquired our measurements adhering to the method as outlined in section 1.2, first running our MilliQ blank several times to ensure consistency, followed by each standard once in order of increasing concentration. This was done in succession using one specific buffer reagent at a time, switching to the next buffer only after all blanks and standards were run. While the standards provided some insight, the blanks were the primary focus for the purpose of comparison. Due to time constraints, quartz-distilled buffer was compared with HPLC buffer on one occasion, and with Optima buffer on a separate occasion. Based on the results, further tests were performed with Optima ammonium hydroxide and optima acetic acid on a third occasion. Given the minor day-to-day variability in baseline absorbance readings, this approach gave adequate basis for comparison between the alternative buffers and our quartz-distilled buffer.

Data Acquisition

In an effort to better understand the uncertainty in our data acquisition strategy we tested two different data acquisition devices (DAQs) to convert our spectrophotometer's analogue signal to a digital one. DAQ A is a LabJack T4, and DAQ B is a Pasco Science Workshop 750. Both DAQs have a bit depth of 12 bits, and we typically run at a 2 Hz sample rate. Both DAQs were physically connected to two screws on the back of the spectrophotometer. The DAQs read and report the voltage difference between the screws, which corresponds to absorbance. The goal of this experiment was to determine the numerical uncertainty in each DAQ and how this might propagate to our final reported values. Using the method described in section 1.2, we ran a full set of known standards and blanks on each DAQ. We then made standard curves with the samples, quantified the blanks, and compared the results of the two DAQs. After obtaining the standard curve for DAQ B we also decided to run the device at higher sample rates than normal to see if this had any effect on our results. In addition, the Pasco Capstone software used for data acquisition with DAQ B has a data smoothing feature that uses a Savitzky–Golay filter which we tested to see if it increased precision.

Rinse pH and Concentration

There was a consistent drop in absorbance directly before the peak: just before each peak there is a small plug of rinse solution in the column that replaces the HCl in the reaction stream. It seemed likely the rinse might be the cause of this drop, potentially due to a different pH or ionic strength in the reaction stream. Efforts were made while making the rinse solution to ensure it achieved the same pH in the reaction stream as the HCl, but perhaps those efforts were insufficient.

In order to investigate this drop in absorbance, the rinse was divided into different concentrations and pHs. The rinse is typically set at a pH of 3-3.5, and a concentration of 3.25 M. The rinse solution was divided into full and half concentrations. Each full and half concentration solution was split into three pHs: 2.11, 3.33, and 3.63. In total there were six bottles: three half concentrations at the three pHs, and three full concentrations at the three pHs. The same seawater was analyzed using each of these rinse solutions, and the peak shape and size compared. Each of these different rinses were analyzed using the same seawater,

Results and Discussion

Blank Measurements

The results of our blank measurement are depicted in Table 1. The overall average blank concentration was 0.53nM ± 0.07 . There was not significant evidence that load time had any effect on measured absorbance. This implies that the MilliQ is largely iron-free, indicating that the blank measurements reflect the iron concentration of the entire preconcentration system.

Additionally, the lack of correlation between load time and absorbance value in our blank allows us to measure our blank at only one load time even if we are running samples at multiple load times. Using the median standard deviation, and assuming that the detection limit is three times standard deviation of all blank runs (.068nm), the detection limit is 0.21 nM

Table 1: Blank concentrations and standard deviations for different load times of MilliQ. Load time does not appear to affect the blank measurement. We estimate our detection limit to be three times the standard deviation of the overall blank: 0.21 nM.

Load Time	Avg (nM)	St. Dev. (nM)
1 Min	0.65	0.26
2 Min	0.51	0.07
4 Min	0.52	0.02
Overall	0.53	0.07

Buffer Grade

We found significant variation in absorbance peak heights while using different quality NH4Ac buffer solutions. Specifically, peak heights of both blanks and standards dramatically increased with the use of either HPLC or Optima grade buffer when compared to quartz-distilled (Figure 3). The average increase in blank peak heights is comparable between HPLC and Optima buffers, which is surprising as Optima grade reagents are specifically made to have extremely low metal concentrations, and we use other optima reagents

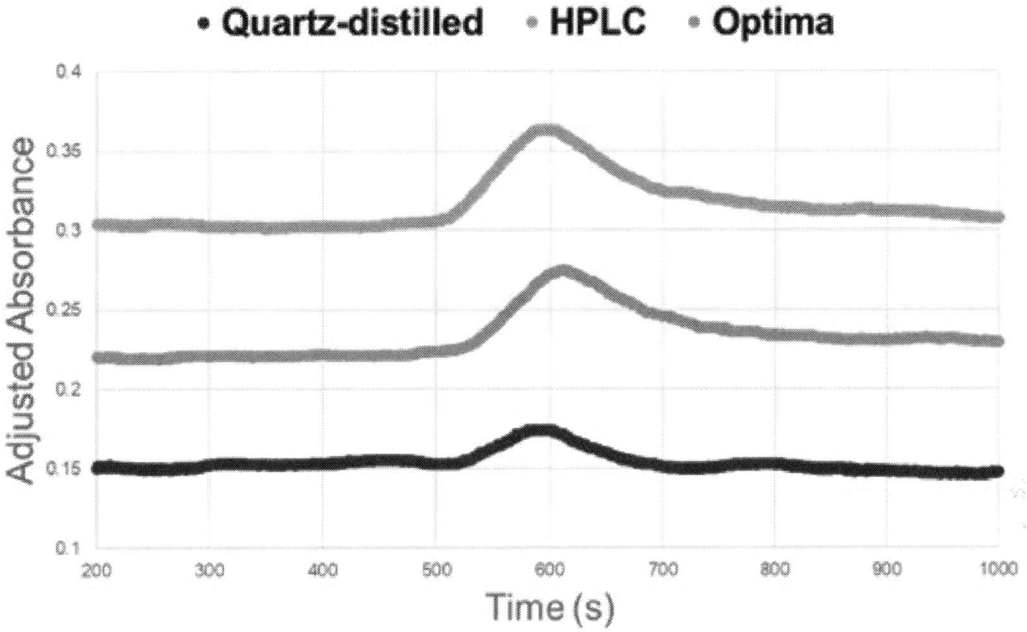

Figure 3: Stacked absorbances of blanks ran with HPLC, Optima and quartz-distilled grade buffer and rinse solutions. Quartz-distilled reagents showed significantly lower blank absorbances with these initial tests.

Table 2: Blank concentrations with varying buffer and rinse solutions. Note that the quartz distilled (Q) solutions result in a substantially lower blank than the HPLC and Optima solutions. As the Optima solution was undoubtedly trace metal clean when purchased, we suspect that the 3 nM blank with the Optima and HPLC solutions reflects something in our preparation of the solutions, rather than actual iron levels.

Buffer & Rinse Solution	Avg. Concentration (nM)	Standard Dev. (nM)
Q	0.50	0.13
HPLC	3.12	0.14
Optima	3.01	0.54

without trouble in other methods in our lab. This increase is reflected in the calculated blank concentrations shown in Table 2. Given that MilliQ should have insignificant iron concentrations, and that the same MilliQ blank solution was used for all experiments, these results suggest larger blanks in using HPLC and Optima reagents in place of quartz-distilled. This was very unexpected, so we thought there might have been contamination while making our saturated solutions from the solid ammonium acetate crystals.

We tried using optima ammonium hydroxide and optima acetic acid to make the rinse and buffer solutions instead. After preparing the rinse and buffer solutions, they were used to measure blanks, standards, and a sample. Testing this

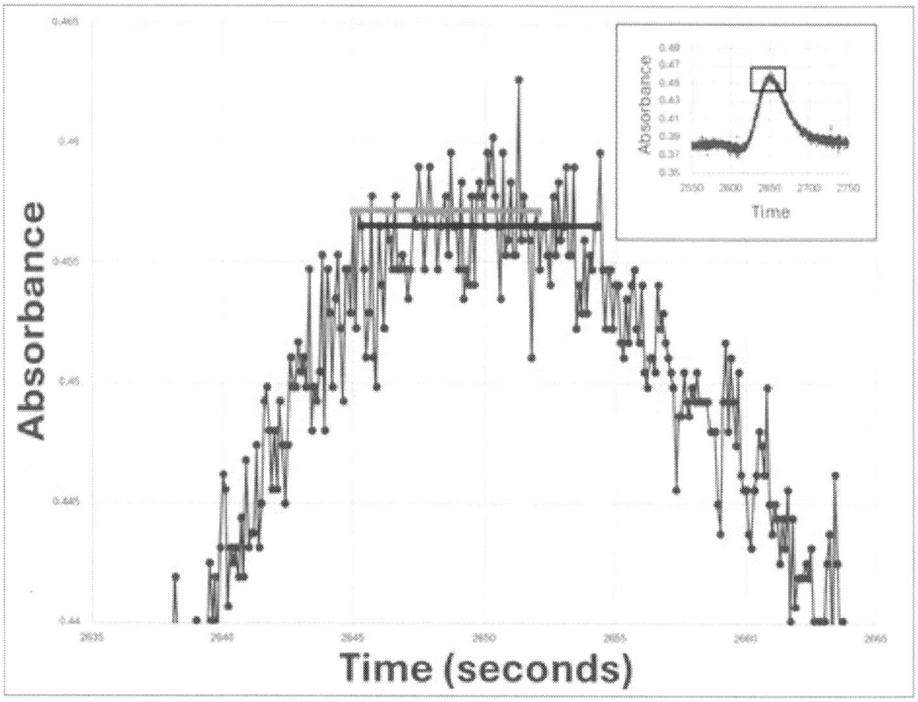

Figure 4: Zoomed in section of absorbance vs time for DAQ B showing the aforementioned data lines. Window in the top right depicts the portion of the peak. The difference between the red and green lines is the minimum absorbance change the DAQ can report. When converted to concentration using our standard addition curve, this corresponds to ~0.03 nM. This reflects limitations in our data sampling abilities.

new buffer resulted in an unusually high baseline and larger than average peaks compared to the quartz-distilled buffer: in other words, these optima reagents behaved in much the same way as when we tried optima (and HPLC-grade) ammonium acetate.

As it is very unlikely that the optima reagents actually have significant quantities (~3 nM, Table 2) of iron, we expect something else is at play here. There may be a concentration difference between our original solutions made from quartz-distilled reagents, and the ones we have tried to make using HPLC and optima crystals and optima saturated solutions. We are not sure why there would be a concentration difference, but that is all we can think of, since we adjusted the pHs to be the same, and the reagents were clean. Further, we have seen previously that lower concentration of rinse solution leads to higher peaks. More work is needed to assess the use of optima reagents in our method.

Data Acquisition

While analyzing the data it became apparent that both DAQs have a lower limit on the data they are able to collect. It seems when data falls between one of these 'limits' it is rounded up or down. This creates the straight lines of data, seen in Figure 4. On both devices the vertical distance between these lines is consistent everywhere. By finding the differences between the y-values of the points on different lines, and using our standard curve to quantify these differences, we were able to find the lower limit on the data we can collect. DAQ A can measure to the nearest 0.033 nM and DAQ B can measure to the nearest 0.028 nM. Given random errors throughout the method, these are essentially the same value, indicating that neither option for the DAQ is strictly better than the other. Based on the detection limit calculated in section 3.1 of 0.21 nM, the limitations on our DAQs are not consequential to error in our overall measurements.

When we ran DAQ B at high sample rates, the

peaks were higher with each increase in sample rate. This makes sense because there is a moment in time when the peak is at a maximum absorbance, and with higher sampling rates of the DAQ, it is more likely that the absorbance will be measured closer to the maximum. Next time, it would be useful to get a standard curve at every sample rate so we can see how accurate these tests are.

Finally, the data smoothing algorithm in Pasco Capstone was also tested, and consistently under approximated the peaks. It seems that the smoothing algorithm should not be used during this data analysis.

Rinse pH and Concentration

Full concentration rinses yielded no visible difference in peak size or change in dip (Figure 5), while half concentrations yielded peaks almost twice as large (Figure 6). Previously our lab group found that insufficient time rinsing the column after the load phase leads to similarly sized peaks. We suspect, therefore, that the problem with the lower concentration rinse solution is that it is similarly not fully rinsing the column. This perhaps leads to high concentrations of salts in the reaction stream, causing the large peak.

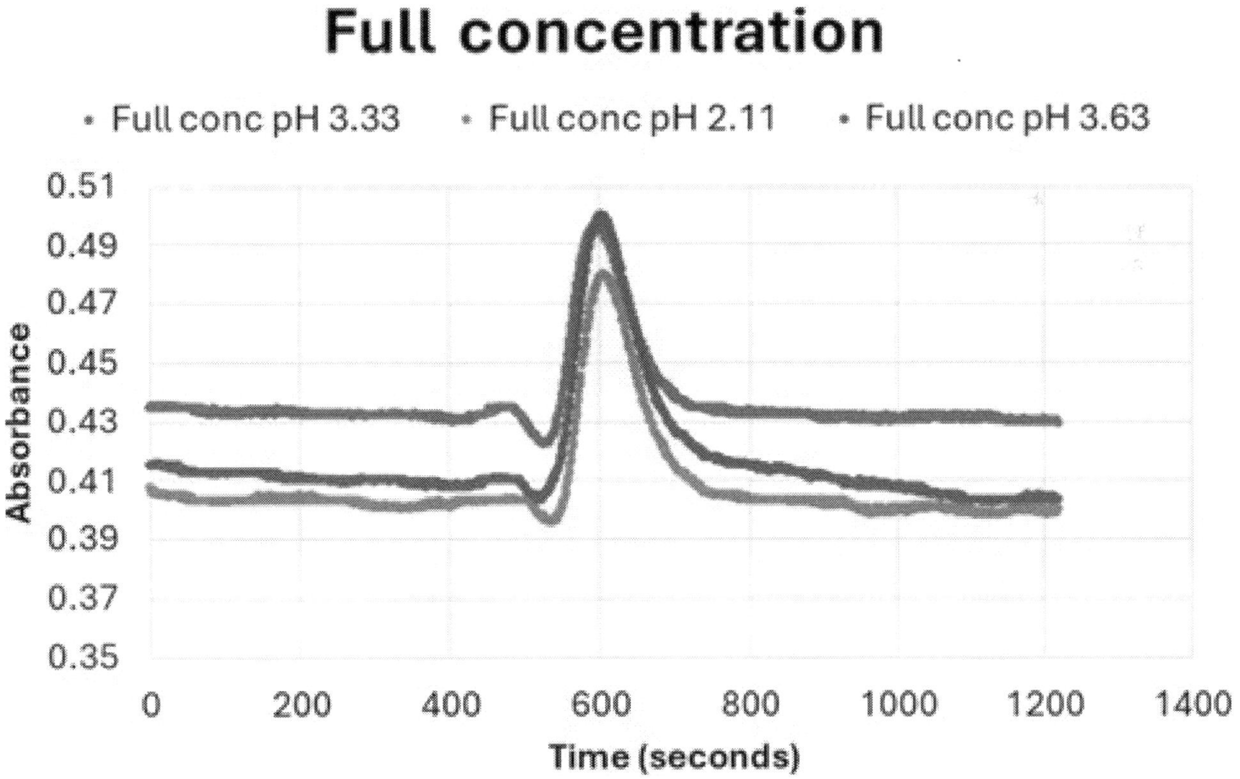

Figure 5: The effect of rinse solution pH on peak shape. Despite our expectations that the dip just before the peak would be affected by rinse solution pH, there did not appear to be any difference in the dip across the three pHs tested. We note that there was a small difference in the baseline between the three runs, however the rinse solution does not contribute to the reaction stream being measured by the spectrophotometer except in the moment just before the peak; further, the baseline routinely shifts throughout the day. Therefore, we do not expect that the baseline shift had to do with the rinse solution.

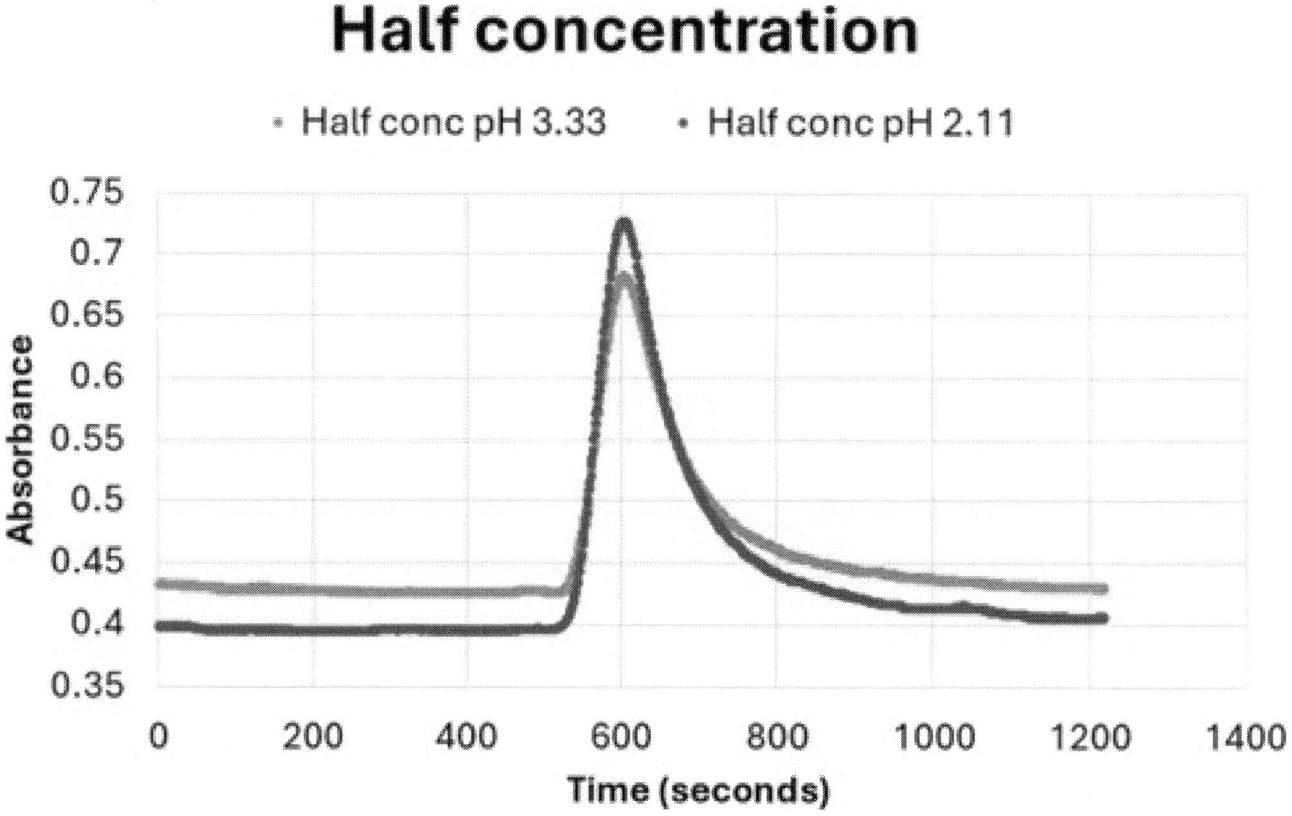

Figure 6: The effect of rinse solution pH on peak shape. Despite our expectations that the dip just before the peak would be affected by rinse solution pH, there did not appear to be any difference in the dip across the three pHs tested. We note that there was a small difference in the baseline between the three runs, however the rinse solution does not contribute to the reaction stream being measured by the spectrophotometer except in the moment just before the peak; further, the baseline routinely shifts throughout the day. Therefore, we do not expect that the baseline shift had to do with the rinse solution.

Change in pH had no outward effect on the dip or peak size (Figures 5&6). This does not explain the dip but does suggest that we no longer need to be as careful with adjusting pH when making the rinse solution.

Conclusions

Our absorbance values were extremely similar between load times, thus making it unnecessary to run blanks at every load time prior to running samples going forward. There must not be significant amounts of iron in our MilliQ, so the absorbance value of the blanks reflects the latent concentration of iron in our preconcentration system. With the default method outlined above in section 1.2, the average concentration of iron in our preconcentration system is 0.53 nM, allowing us to compare all further adjustments to the method to this value. In the future, we will double-check our findings here by making a set of standards in MilliQ rather than seawater. It is possible that the salts in the seawater change the sensitivity of the method, in which case the standard curve that applies to the MilliQ blanks would be different than what applies to the standards. Doing this additional test would add confidence to the accuracy of our blank measurements reported here.

Implausibly high (3 nM) blanks measured with both the HPLC and particularly the optima grade ammonium acetate motivated us to purchase optima glacial acetic acid and optima ammonium hydroxide to reattempt these buffer tests: in other words, starting with optima solutions rather than optima solids. The optima glacial acetic acid/ammonium hydroxide buffer results were also implausibly high, with an uncharacteristically high baseline and larger than average peaks. The rinse and buffer solutions made with our limited amount of quartz-distilled ammonium acetate (which give us the reasonable blanks in Table 1), were the same pH and were intended to be the same concentration as the rinse and buffer we made from optima reagents in this work. Therefore, it is surprising that they behave so differently. More work is needed to figure out what the difference is with our new efforts. One observation is that the pH of our new ammonium acetate solutions are different (initially, before pH adjustment) than we get with our quartz distilled crystals: another line of evidence for there being some key difference between the solutions. This could potentially be a result of different ratios of ammonium and acetate in our quartz distilled crystals.

Our research into data acquisition revealed that both DAQs we have used in the lab are capable of taking data with adequately low uncertainty, and neither DAQ is strictly better than the other. DAQ A can measure to the nearest 0.033 nM and DAQ B can measure to the nearest 0.028 nM (Figure 4). Although these uncertainties are low, it would be ideal for them to be as small as possible. Future research in this area will include standard curves at higher sample rates to see if they are more accurate and looking into how to reduce our data acquisition uncertainty even further, perhaps by oversampling and averaging data points.

The rinse test revealed that pH had a negligible effect on peak shape, but concentration of the rinse solution drastically impacted peak size. This reveals that there is less need for as much vigilance in ensuring the reagents were the correct pH. The higher baseline and peaks that were produced by the half concentration rinses further confirmed the necessity of the correct concentration. However, neither of our tests revealed the cause of the pre-peak dip; more work is needed to address the dip problem.

Acknowledgements: The authors would like to thank Michael Barth for his help editing this manuscript and Tyler Hooker for helping with data acquisition.

References

Biller, D. V., Coale, T. H., Till, R. C., Smith, G. J., & Bruland, K. W. (2013). Coastal iron and nitrate distributions during the spring and summer upwelling season in the central California Current upwelling regime. Continental Shelf Research, 66, 58–72. https://doi.org/10.1016/j.csr.2013.07.003

Lohan, M. C., Aguilar-Islas, A. M., & Bruland, K. W. (2006). Direct determination of iron in acidified (pH 1.7) seawater samples by flow injection analysis with catalytic spectrophotometric detection: Application and intercomparison. Limnology and Oceanography: Methods, 4(6), 164–171. https://doi.org/10.4319/lom.2006.4.164

Moore, J. K., Doney, S. C., & Lindsay, K. (2004).

Upper ocean ecosystem dynamics and iron cycling in a global three-dimensional model. Global Biogeochemical Cycles, 18(4). https://doi.org/10.1029/2004GB002220

IdeaFest
Journal

The Art of Martin Wong, and the Traditions He Drew From

Janie Mendosa
Department of Art & Film, Cal Poly Humboldt

Even with his formal art training, prolific painting practice, and charming personality that endeared him to many social groups, Martin Wong's art and life existed on the margins. He wrote to his long-time friend, Gary Ware, in 1967 that he was "always the outsider wandering the edges" (Yau, 2023). Wong self-identified as "Chino-Latino," although he spoke neither Chinese nor Spanish; he was also a gay man who was witness to, and eventually tragically succumbed to, the AIDS crisis. As a street portraitist "pencil for hire" in Old Town Eureka, a hippie during the Summer of Love in San Francisco, an urban cowboy in pre-gentrified New York City, and a "tourist" in Chinatown and Puerto Rican neighborhoods, his art was esoteric and visionary, yet always based in observation. Throughout his life, Wong voraciously collected art, refusing to discriminate between "high" and "low" quality. He kept an original Mondrian drawing across his sink in a run-down New York walk-up. Wong's apartment was filled with "Chinese ceramics, the paintings of Thomas Eakins and Winslow Homer, calligraphy, archers' thumb rings, children's lunchboxes, Mickeys and Minnies and Donalds…" (Ault, 2013). Similarly, his artwork eludes easy categorization. His large scale oil and acrylic paintings often included densely layered motifs of brick walls, "sign language, astronomical constellations, graffiti, Loisaida [the predominantly Puerto Rican neighborhood, "Lower East Side"], the writing and person of Miguel Piñero [a Puerto Rican playwright, Wong's friend, occasional collaborator, and lover], men in prison, firefighters, Chinatown…" (Ault, 2013). Although Wong had a unique and inimitable visual language that he developed, he drew from numerous disparate Western and Asian traditions, including modernist painting traditions, Pan-Asian iconography, and the graffiti art tradition.

Wong was never closely associated with any contemporary post-modern movements. Instead, he has been described as a social realist painter, similar to the turn of the century Ashcan School of art (Blinderman, 2024). The "Ashcan School" was a small group of artists including Robert Henri and John Sloan, so named because they focused on depicting urban scenes in the lower classes in New York City (McNamara, 2007). Most of the artists from this group began as newspaper illustrators who honed their observational drawing skills similarly to Wong. Both Wong and the artists of Ashcan School depicted unsettling, hyper-local scenes of New York City's poorest neighborhoods during a period of radical change. Although all of these artists were documentarians of their time, they declined to engage in overt political commentary (Homer, 1969). Despite his similarities to the Ashcan School, Wong's primary influences were WPA (Works Progress Administration) paintings and social realist work from the thirties and forties, including Diego Rivera and Frida Kahlo (Machida, 2022). Dong Kingman's early WPA work of Chinatown is another social realist whom Wong cited as an early role model. He also was inspired by Cubist artists such as Picasso and Braque, not due to their style, but because after he saw photographs of Picasso's environment, he believed that the artists "'were mostly drawing [Cubism] out of their direct experience rather than an artistic motif'… Wong decided that everything [he] use[s] in… the Lower East Side paintings, would be things [he] actually saw rather than something [he] had picked up" (Machida, 2022). His observational painterly style was anachronistic for his time, unlike the stark and minimalist Neo-Geo [Neo-Geometric Conceptualism], Abstract Expressionist, and Pop Art movements that were trendy commodities at the time (Blinderman, 2024).

In many of Martin Wong's works, there are references to Asian iconography and ideas, not just from his Chinese background, but general Pan-Asian visual traditions. In 1971, he traveled to Nepal, India, and Afghanistan, and was fascinated by the tantric and esoteric imagery there (Johnson, 2022). During the early 1970s, he was obsessed with the checkerboard-like pattern of Balinese *saput poleng* textiles. The black and white squares represent the balance of opposing forces in Hindu tradition (Johnson, 2022). Wong's final piece *Did I Ever Have a Chance?* depicts a blue skinned Patty Hearst as Kali, the Hindu goddess of death, brandishing a rifle and wearing a *mundamala*, a traditional garland of skulls (Johnson, 2022). While Wong references Asian tradition, it is often wrapped in layers of translations and nods to pop culture. For example, *Heaven* (1988), a six-foot diameter canvas circle with painted bricks and a black hole in the middle, references the form of ancient Chinese jade discs (*bi*), yet still is also a brick wall with a gloryhole (Ault, 2015)—a wink at his queer identity and urban environment. Bruce Lee, San Francisco-born Hong Kong-based martial artist and movie star, is a repeating figure in Wong's work. Wong's relationship with his Chinese heritage was complex. Although his mother, Florence, spent her childhood and young adult life in China, and his stepfather was mixed Mexican and Chinese, he didn't fit in with the larger Chinese-Asian community. Although he grew up near San Francisco's Chinatown, and began his lifelong interest with Asian antiques in childhood, Wong stated, "I grew up in Little Italy and sometimes I feel almost as much Italian as Chinese" (Machida, 2022). Later, he said, "I feel as much a hippie as a Chinese American," when referring to his formative years in the "overwhelmingly white" counterculture movement in San Francisco and experiencing the lasting effects of LSD (Johnson, 2022). As early as the 1960s, Wong experimented with the format of long Chinese scrolls, creating vertical scrolls featuring hard to decipher poetry. He returned to this format later in his Chinatown series, such as *Grant Avenue, San Francisco* (1992) and *Mei Lang-Fang* (1992). In his interview with Machida, he states, "Like a lot of Caucasian artists, they look back to their painting traditions of like Velasquez or Rembrandt, myself as a painter, when I look back on a tradition it's to Lamqua, to the Canton painting school." Lamqua (ca. 1802–60) was a masterful portraiture artist who was one of the first Chinese painters to use Western oil painting techniques. Through Lamqua, Wong identifies with and is interested in the merging of cultures, rather than the formalism of traditional aristocratic East Asian ink painting. Machida posits that he was also inspired by Cantonese trade paintings created by apprentices for commercial purposes that show everyday scenes such as storefronts, workshops, and trades.

A common motif in Wong's work is hands using ASL fingerspelling as a script to create difficult to decipher glyphs. Not only did he draw inspiration from the deaf community, he also drew from two separate Asian traditions: Chinese landscape paintings, and Kufic text. In a 1984 interview with Yasmin Ramírez, Wong states, "Basically, I'm a Chinese landscape painter. If you look at all the Chinese landscapes in the museum, they have writing in the sky. They write a poem in the sky, and I do that, too." His vertical compositions also harken back to scrolls of East Asia. During his travels to Afghanistan, he picked up a ninth century Kufic manuscript that specifically inspired him: "When I elongate the fingers in my paintings, a lot of times it relates to the stylization that happened in Kufic, when they tried to use it in architecture" (Ault, 2013). As an Asian American, Wong had the creative liberty to draw from many rich sources.

During his time in New York, Wong became a graffiti collector, buying artwork from artists like Lee Quiñones, Daze, Sharp, Keith Haring, and Lady Pink. Although Wong never was a graffiti artist himself, he was an avid supporter of the movement and included many references to the newly blossoming tradition in his work. In 1989, he opened a short-lived gallery dedicated to graffiti. In 1994, he entrusted the 300-piece collection to the Museum of the City of New York. The tradition of writing on the walls is one of the oldest in history, but the modern graffiti movement most likely started in Philadelphia in the late 60s (Corcoran, 2013). It spread shortly to New York City where it exploded in popularity. Martin Wong's most famous work, *Attorney Street (Handball Court with Autobiographical Poem by Piñero)* (1982–84), is a collaborative piece Piñero commissioned from Wong to depict the neighborhood handball court with graffiti made by Little Ivan (Ivan Torres). Around this time, Wong worked at Pearl Paint in the canvas department, and befriended the young artists (Cocoran, 2013). In Wong's painting *Sharp Paints a Picture (Collaboration with Sharp)* (1997–98), he depicts a shirtless SHARP (Aaron Goodstone) wearing a respirator, clearly in the middle of his criminal, or artistic, act. As an outsider to the art world himself, and a person who thrived on transgression, Wong must have seen himself in the artist. At the same time, the collector in him also enjoyed his outsider status to the graffiti tradition. "The fact that I have never picked up a spray can, to me it makes spray painting a lot more exotic as a collector" (Machida, 2022). Even though he never was a graffiti artist, Wong was drawn to engage in this tradition through monetary support as a collector, and cultural influence as a curator. As an artist, his collaborative work with graffiti artists showcases the influence the culture had on his aesthetic vision.

Martin Wong was a singular figure in the art world, drawing from diverse influences including Pan-Asian traditions, Western modernism, social realism, and the emerging graffiti movement. Throughout his career, Wong was able to masterfully blend social documentary, personal symbology, and cultural references to make art that speaks to the complexity of his lived experience and multifaceted identity as a queer Chinese man. Although in his short lifetime he was underappreciated, he chose authenticity to himself and his interests over what was selling on the art market, or even what was "politically correct" at the time.

References

Ault, Julie. "Martin Wong Was Here." In Martin Wong: Human Instamatic, 83–101, p. 96, London: Black Dog Publishing, 2015.

Ault, Julie. "Some Places it will always be Eureka and in Eureka it will always be Valentine's Day." In *I.M.U.U.R.2*, edited by Julie Ault, Peter Knes Heinz, Danh Voˉ, Christopher Müller, and Daniel Buchholz, n.p. K.ln, Germany: Galerie Buchholz, 2013, https://cdn.contemporaryartlibrary.org/store/doc/5086/docfile/original-9769c45a7b46f658e647eb142164c43e.pdf.

Blinderman, Barry. "Martin Wong's Chains of Desire." *Evergreen Review*. Accessed 18 Dec. 2024, https://evergreenreview.com/read/martin-wongs-chains-of-desire/.

Corcoran, Sean, and Carlo McCormick. *City as Canvas: New York City Graffiti from the Martin Wong Collection*. New York, NY: Skira Rizzoli, Museum of the City of New York, 2013.

Homer, William Innes. "The Exhibition of 'The Eight': Its History and Significance." *American Art Journal 1*, no. 1, 1969, pp. 53–64, https://doi.org/10.2307/1593854.

Johnson, Mark Dean. "Pieces of the Puzzle: Martin Wong's Multiplicity." *Martin Wong Catalogue Raisonné – Spotlight*, Stanford: Stanford Libraries, 25 July 2022, https://exhibits.stanford.edu/martin-wong/feature/pieces-of-the-puzzle-martin-wong-s-multiplicity-mark-dean-johnson.

Machida, Margo. "Martin Wong: Renewing Our Conversation." *Martin Wong Catalogue Raisonné – Spotlight*, Stanford: Stanford Libraries, 11 July 2022, https://exhibits.stanford.edu/martin-wong/feature/martin-wong-renewing-our-conversation-margo-machida.

Machida, Margo, and Martin Wong. "1989 Martin Wong: Interview with Margo Machida." *Martin Wong Catalogue Raisonné – Spotlight*, Stanford, 18 July 2022, https://exhibits.stanford.edu/martin-wong/feature/1989-martin-wong-interview-with-margo-machida.

McNamara, Kevin R. "Review of Up Close and Impersonal: Ashcan Art, City Life, and Democratic Politics, by Rebecca Zurier." *The Journal of the Gilded Age and Progressive Era 6*, no. 4, 2007, pp. 476–78, http://www.jstor.org/stable/25144502.

Ramírez, Yasmin. "Writing in the Sky: An Interview with Martin Wong." *East Village Eye*, Oct. 1984.

Yau, John. "Martin Wong, the Perennial Outsider, Answers Back." *Art Basel*, 21 June 2023, https://www.artbasel.com/stories/martin-wong-queerness-john-yau.